Measuring Operations Performance

More information about this series at http://www.springer.com/series/13500

Andrea Chiarini
Editor

Sustainable Operations Management

Advances in Strategy and Methodology

Springer

Editor
Andrea Chiarini
Department of Economics and Management
University of Ferrara
Ferrara
Italy

ISSN 2363-9970 ISSN 2363-9989 (electronic)
Measuring Operations Performance
ISBN 978-3-319-14001-8 ISBN 978-3-319-14002-5 (eBook)
DOI 10.1007/978-3-319-14002-5

Library of Congress Control Number: 2014957399

Springer Cham Heidelberg New York Dordrecht London

Printed on acid-free paper

Springer International Publishing AG Switzerland is part of Springer Science+Business Media (www.springer.com)

Contents

Marketing Strategy, Strategic Planning and Corporate Social Responsibility: An Exploratory Research

Andrea Chiarini

Abstract In the last decades companies have started embracing Corporate Social Responsibility (CSR) for their Marketing Strategy (MS). Ethical scandals as well as financial scandals linked to companies and their brands have jeopardized several companies and consumers have even started boycotting purchases. Moreover, starting from the 1980s, mass marketing has been transformed into mass customisation and markets have been divided into micro-markets and even individual markets. In such a scenario customers and markets have become very interested in CSR impacts of the products/services sell by the companies and customers often compare products using this way of thinking. As a consequence, companies have to integrate their strategic marketing initiatives with issues such as environmental management, health and safety for workers, social responsibility and have to take into account how this can affect sales. To the purpose, this research wants to investigate what are the methodologies for strategic planning and MS and subsequently what are the CSR issues related to it. The research is based on a case study of four European large manufacturing companies and demonstrates how MS can change depending on the kind of market and product, even if the budget for MS is similar.

Keywords Marketing strategy · CSR · European manufacturing · Qualitative inquiry

1 Introduction

In the last decades companies have been struggling with a relevant change of scenario concerning their strategies. Companies have been facing new markets, different consumers and issues such as greening the processes and social responsibility; thence

A. Chiarini (✉)
Department of Economics and Management, University of Ferrara, Ferrara, Italy
e-mail: andrea.chiarini@unife.it

A. Chiarini (ed.), *Sustainable Operations Management*,
Measuring Operations Performance, DOI 10.1007/978-3-319-14002-5_1

Corporate Social Responsibility (CSR) in general has become over time more and more important.

Since 1990s many scholars have focused their papers on CSR and how CSR can pay off the investments (Burke and Logsdon 1996). Strategies for social responsibility are moving companies from philanthropic local initiatives to world-wide projects and Marketing Strategy (MS) for all the markets (Bhattacharya et al. 2004). In fact, companies in some way have been forced to embrace CSR for their Marketing Strategy (MS). Ethical scandals as well as financial scandals bound to companies and brands have jeopardized several companies (Baccarani 2008) and consumers have even started boycotting these companies. Moreover, starting from the 1980s, mass marketing has been transformed into mass customisation and markets have been segmented in micro-markets and even individual markets (Kotler 1989). In such a scenario some customers and markets are very interested in CSR impacts of the products/services sell by the companies (Sen and Bhattacharya 2001; Klein and Dawar 2004; Nan and Heo 2007) and can compare products using this way of thinking. As a consequence, companies have to integrate their strategic marketing initiatives with issues such as environmental management, health and safety for workers, social responsibility and have to take into account how this can affect sales.

To the purpose, this research wants to investigate what are the methodologies for strategic planning and MS and subsequently what are the CSR issues related to it. According to literature review section, several authors have dealt with methodologies and principles for Strategic Planning and MS but there is a lack concerning how to integrate CSR issues with theses methodologies and principles. For instance, it is not that clear what the stages of this integration are, what specific methodologies are typically employed, what kind of deployment is managed and what relationships with the customers and the markets are depending on the kind of product.

As to as, a multiple case study based on four European large companies has been carried out, trying to investigate the above mentioned issues. The research is for its nature exploratory and from the research new theory will emerge in an inductive way. Four European large companies have been observed and information about CSR and MS has been collected, coded, analysed and discussed in order to get a theoretical framework.

Therefore the remainder of this paper is structured as follows. Section 2 deals with Strategic Planning, MS and their principles and consolidated methodologies. Section 3 reviews some relevant literature concerning Strategic Planning, MS and CSR, trying to understand what other authors have investigated and claimed so far. Section 4 deals with the research methodology, its characteristics and its limitations. The section also explains what kind of companies has been chosen for the research and the fundamental organisational processes of these companies. Section 5 analyses and discusses the results from the observation, coding the results for achieving a theoretical framework. Lastly, the conclusions section summarises the findings of the research, what is the fundamental novelty and the limitations of the research. Moreover some avenues for further research on the subject are presented.

2 Strategic Planning Process

A strategic plan is a methodology to guide the work of managers and the organisation as a whole, maintaining them focused on long term objectives, vision and mission and their deployment. It also aids decisions about the allocation of companies' resources such as human, technological and financial resources. Strategic planning can be seen as a rational, formalised approach led by decision makers to head organisations toward strategic decisions (Mintzberg 1994). According to Mintzberg (1994), strategic planning can sometimes fail because its assumptions are completely disconnected from the way persons and organisations learn and make decisions, including CSR issues. As Mintzberg wrote (p. 321): "Thus we arrive at the planning school's grand fallacy: Because analysis is not synthesis, strategic planning is not strategy formation".

Having underlined this relevant aspect of strategic planning, the process can be then divided into specific stages and processes. According to Hillestad and Berkowitz (2012) there is a fundamental relationship between company strategy and MS. Figure 1 below shows the loop which links the strategic planning process to MS.

Thence, MS has a direct link with Mission and Vision statements as well as strategic goals and objectives which are usually shared through a business plan (Andrews 1997). In this way the deploying flow process starts with the Mission, Vision and strategic goals and ends with a periodical review of the marketing plans and an eventual correction of the flow. Figure 2 synthetically shows this flow and the main stages. The figure reflects what other authors have claimed so far (Jocumsen 2004; Wilson and Gilligan 2005; Cravens and Piercy 2008; Lord and Velez 2013).

Within the second stage there is the fundamental SWAT (Strengths-Weaknesses-Opportunities-Threats) process (Lymbersky 2008) which leads the company to evaluate the strengths, weaknesses, opportunities and threats involved in defining

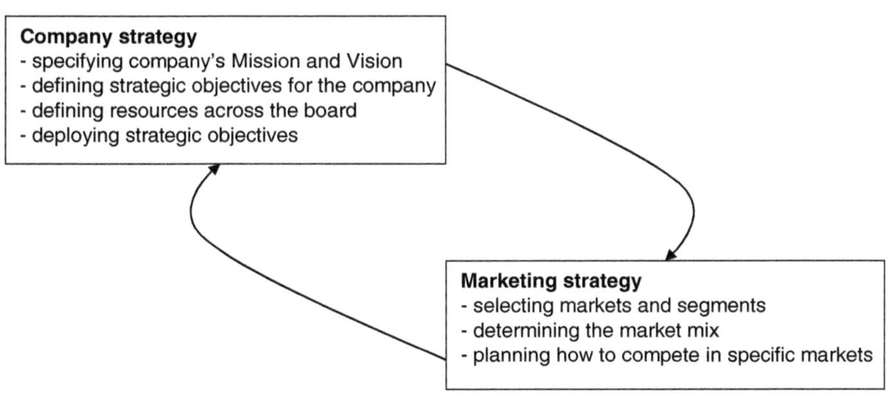

Fig. 1 Relationship between company and marketing strategy

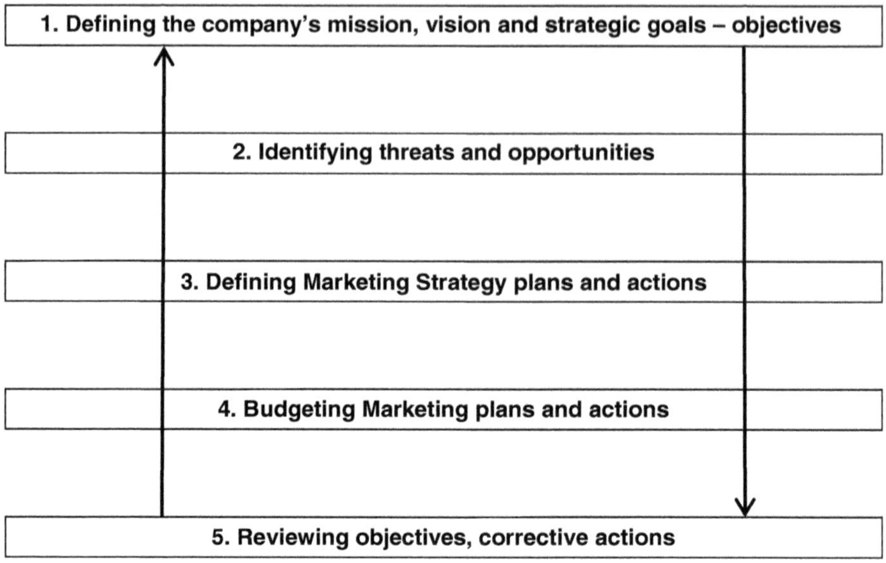

Fig. 2 Deployment from long–period strategies to the periodical reviewing process

market strategies (Ronald 2012). The five stages shown by Fig. 2 have been used in this paper for analysing and discussing how the companies chosen for the explanatory research are integrating their CSR strategies.

3 Literature Review

There is plenty of literature dedicated to SM; for the scope of this research just literature which has a link with CSR has been taken into account. The first interesting papers dedicated to the integration between SM and CSR dated back to 1990s. McDaniel and Rylander (1993) wrote a paper in which they argued that (1993, p. 4): "… green marketing is taking shape as one of the key business strategies of the future, and that the increasing environmental consciousness makes it incumbent on consumer marketers not just to respond to it, but to lead the way in environmental programs. Sets forth the need and recommendations for incorporating environmental concerns into strategic marketing planning of the organisation".

McDaniel and Rylander were not the only pioneers in that period. Similar issues relating to the integration between green marketing and strategies can be also found in Polonsky (1994), Mendleson and Polonsky (1995) and Straughan and Roberts (1999).

Other author started investigating CSR in general and its relationships with strategies, arguing that CSR pays off the investments when well implemented

(Burke and Logsdon 1996). Quantitative inquiries carried out by Brown and Dacin (1997) and Brown (1998) demonstrated that CSR can also positively affect, directly or indirectly, consumers' responses.

Freeman (2001) wrote a paper, taking up again later by Lantos (2001), in which he claimed that CSR has to become a strategic issue and cannot be just an altruistic activity. Both the authors concluded that CSR is good for companies and business, advising that marketing takes a lead role in strategic CSR activities.

The 2000s are consequently dominated by papers where scholars investigated the role of marketing and MS as a vehicle for CSR and vice versa. McWilliams and Siegel (2001) argued that for satisfying the developing demand of CSR from various markets and customers, companies have to put on the market products which include specific CSR characteristics. MS becomes in this way important for advertising these particular CSR products as well as a segmentation of markets is important for analysing how to satisfy the different stakeholders. Du et al. (2010) underlined the power of communication plans for boosting companies' messages to customers; message content and communication channels are the specific factors that influence the effectiveness of CSR communication.

Piercy and Lane (2009) in their research concluded how relatively little attention has been devoted to the link between CSR and strategic marketing. According to the authors there is a need for a defined value proposition that emphasizes how a company's CSR strategy adds value for the customer. Moreover they proposed a new management agenda for marketing strategy that examines CSR opportunities and risks. However marketing managers need to analyse carefully the market context as a whole before implementing CSR strategies. For instance Bhattacharya and Sen (2004, p. 9) demonstrated that it is important for managers to consider CSR strategies in the light of the firm's corporate abilities. In fact, it seems that less innovative companies may be better off financially by avoiding CSR action.

To recap, literature review seems having demonstrated the need of MS for CSR and even that CSR can, in many circumstances, really help companies in competing and increasing financial performances. However it is not that clear in what way this can be done, and what kind of methodologies and principles can be applied for aiming at this goal.

4 Methodology

Methodology is mainly based on case studies carried out in four European large manufacturing companies. Inside qualitative inquiries and research design, the case study represents one of the most used methods especially in the management sector. In point of fact, some researchers (Stake 1995) argue that a case study is an object of analysis more than a method. Other authors point out that it is a typical qualitative method of inquiry used as an exploration of a 'bounded system' (Merriam 1988). According to Creswell (1998), time and place are the boundaries of the case study and the researcher is supposed to use multiple sources of information.

When using a case study a researcher first has to take into serious account if the case can be single or multiple, multi-sited or within-site, focused on a case or on an issue (Yin 1994). According to Dogan and Pelassy (1990), the use of case studies cannot generate theories in a definitive way. Case studies must be linked to a hypothesis that can then be followed by a deductive path. The use of case studies in this research is justified by the facts that:

- It is possible to obtain outcomes that can be compared with other similar case studies. As a matter of fact, in the following section it will be demonstrated how the outcomes of an interview with a senior manager of a company were compared with the outcomes of other three interviewed managers.
- Case studies typically lead to the use of different methods to collect data such as interviews, database, questionnaires and observations. Data can be both qualitative and quantitative, even though only qualitative evidence is analysed in this exploratory research.

Case studies by means of qualitative inductive research have received many criticisms. Anyhow some researchers state that, probably, no predictive theory exists in social studies or management. Social studies and management have never led to the generation of context-independent theories and can therefore only take context-dependent knowledge into account. Campbell (1979) is one of the researchers who support these statements. A quotation from his 1979 (p. 126) book can surely better explain his position: "After all, man is, in his ordinary way, a very competent knower, and qualitative common-sense knowing is not replaced by quantitative knowing. This is not to say that such common sense naturalistic observation is objective, dependable, or unbiased. But it is all that we have. It is the only route to knowledge… fallible and biased thought it be".

Yin (1994) defines case study research as an empirical inquiry that investigates a contemporary phenomenon within its real-life context. Thus, case study research becomes particularly important when the boundaries between phenomenon and context are not clearly evident and multiple sources of evidence are used.

In order to not introduce bias the four companies which represent the case study were chosen with different characteristics in terms of marketing approach and kind of product. Indeed the four companies, hereinafter named A, B, C and D, belong to the same geographic area but they sell their product in different kinds of markets. Furthermore the companies manufacture standard products and even more innovative products. Figure 3 shows the strategic approach to the markets related to the four different companies. In particular company A typically sells standard products for rich markets. In greater detail A sells power plugs for the European and North-American markets. B sells innovative products for rich markets, in particular wind turbines for the European market. While C manufactures standard gas valves for some African countries and D manufactures medical devices for African and Middle-East countries. Obviously, the four companies have been chosen for their well-known commitment and initiatives to CSR and they have similar turnovers.

The researcher has gathered data and information as a neutral observer.

Fig. 3 Marketing strategy approach of the four companies

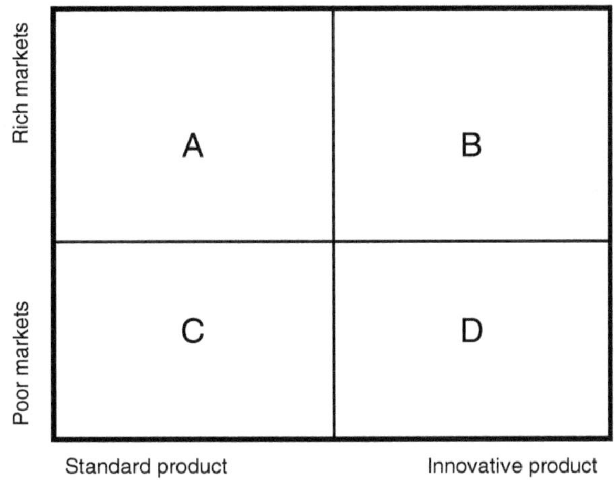

The primary scope of observation is to observe participants in as natural a setting as possible (Pearsall 1995). Observation has been carried out within a group of senior managers involved in MS, looking at and discussing their documents and results.

Each company clearly shows a different way and culture of managing matters such as MS, CSR and relationships with stakeholders. The observation has been led in an ethnographic way and direct experience has been chosen (O'Reilly 2005) within a team in which the group of participants were observed first and then the findings of the observation were discussed with them. The ethnographer has to introduce himself or herself in an unfamiliar territory (Denzin 2000).

By conducting observation, the researcher learns about what is going on at the same time as he or she is building strong and informative relationships with the group. Observation relies heavily on the researcher's subjective understanding of research situations and because of this the researcher must be aware that it requires a careful balance. The researcher led the observation being both a participant and an observer. Indeed the researcher acted as a consultant giving details about how to manage the MS process. The researcher did not interfere with the team dynamics. A mistake that the researcher tried to avoid was to take notes as the dominant part of participant observation and this was mainly due to the researcher also acting as observer-researcher and consultant at the same time.

Data from the observation has been gathered writing down some notes and recording discussions with the senior managers and the team's members dedicate to MS. The practice of coding data (Lofland and Lofland 1995) was used to assign labels to classify and assign meaning to parts of the information.

5 Discussion of the Theoretical Results

5.1 Mission, Vision and Strategic Goals

MS starts its flow with the Mission and Vision statement as well as the definition of the strategic goals/objectives for the company. All the four companies changed in the past their Missions and Visions in order to communicate to all the stakeholders their new CSR aim. According to the senior managers who belong to the four companies, they had to change these strategic documents in particular because they are public and disclosed to anyone; thus they represent the first tangible sign of this transformation. One manager declared that if you want to use CSR leverage for you marketing strategy surely you have to manage a communication plan for stakeholders and markets in general. And the best way to do this is your Mission and Vision. This is the reason why all the four companies changed their Mission-Vision.

Anyhow, looking at the strategic goals and objectives, usually written in the business plans and visions documents, something different depending on the company has come out. Company C which usually sells standard products to poor markets seems to prefer CSR external goals and objectives rather than internal ones. Indeed its goals are more linked to increase customer's satisfaction and reputation as an ethical company dedicated to CSR initiatives as well as to be a company which does not pollute or exploit resources. According to some managers who belong to company C, exploitation of resources and environmental management are a fundamental issue for developing countries. Customers and consumers in such markets want companies, especially world-wide companies, to not have a strategy of just considering them as an underdeveloped economy. They are suspicious and tend to consider world-wide companies like companies more devoted to profit than other issues. This aspect has been also confirmed by managers who belong to company D, which sells innovative products but once more to poor markets.

On the other hand, company A and company B which are more focused on rich markets such as Europe and North America have established strategic objectives which aim to internal issues of the company. In this sense objectives related to staff's satisfaction, health and safety and social responsibility towards employees are preferred. The reasons of this choice are linked to the fact that A and B represent European companies which sell their products to the same area of belonging or similar. According to a marketing senior manager, consumers tend to identify themselves with companies' employees and they buy products only if the company fosters CSR especially towards its own employees. Moreover, according to another senior manager, in advanced and rich markets company's compliance to issues such as environment, exploitation of resources as well as the pursuit of positive impacts on society are taken for granted by consumers.

As a common strategy all the four companies have defined strategic goals associated to compliance with laws for avoiding accounting frauds. After the well-known scandals occurred in the last decade, consumers do not want to spend their money on products manufactured by this kind of companies, irrespective of the kind of product or market.

5.2 Identifying Threats and Opportunities

All the four companies are used to implementing a SWAT analysis in order to define MS for their products. Threats and opportunities are typically analysed and managed on the base of what has been discussed in the previous section. In particular threats and opportunities are different depending on the market rather than the kind of product.

Starting from threats, companies C and D which sell products in poor market are very careful to everything can be linked with exploitation of resources and pollution. Possible environmental incidents which have hit the headlines are considered the most terrible threats for the company and its products in the emergent markets. In this case, according to the senior managers, a particular communication plan has to be developed specifically for those markets to climb back on the top. On the other hand, as an opportunity, companies C and D surely promote every kind of initiative and project which has had a positive impact on the environment and the consumption of resources, especially natural resources.

Instead, companies A and B, specialised in richer markets, consider as a threat each scandal related to health and safety issues and social responsibility towards employees. On the other hand every achievement which increases the satisfaction of the stakeholder employees is appropriately promoted.

For all the four companies financial scandals are considered one of the most important threats whichever is the market.

5.3 Defining Marketing Strategy Plans and Actions

Observing how the managers of the companies deal with their marketing strategy plans and actions, a common flow can be sketched. According to all the managers usually there are four main steps as described in Fig. 4.

Targeting is the activity where the four companies ask themselves who are their customers. This derives from the strategies discussed in Sect. 5.1 and the quadrant in Fig. 3. However, customers have to be further segmented based on a number of factors. Companies C and D are used, for example, to dividing customers into dealers and industrial companies. Especially for industrial companies, marketing efforts in terms of CSR are oriented towards companies which have got an environmental management certification (i.e. ISO 14001) or have publicly demonstrated their commitment towards CSR, for instance issuing a social balance sheet or introducing a social and ethical performance measurement system. Marketing and

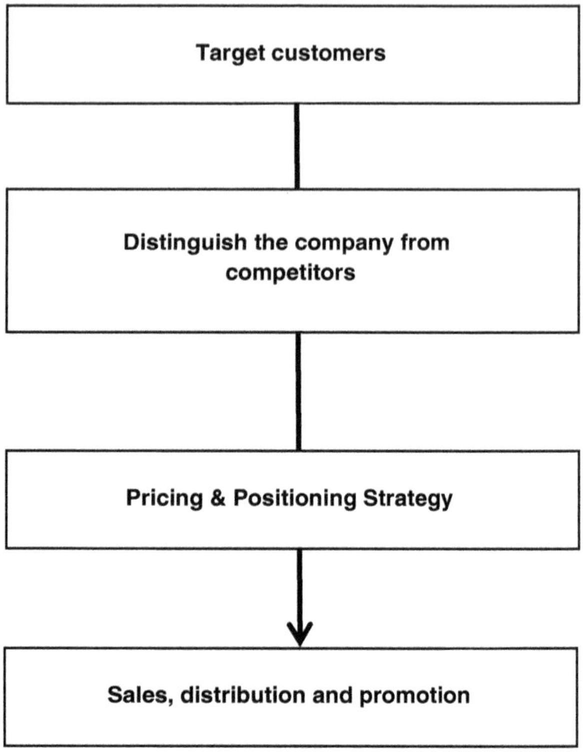

Fig. 4 Typical marketing strategy planning stages

sales managers of companies C and D usually take care of these industrial customers giving them more technical and detailed information about environmental initiatives of the company and product environmental characteristics; for instance ISO 14001 certification, compliance with environmental laws and regulations, recyclability of the products, etc. Above all, customers have to perceive the difference between products from companies C and D and from their competitors from an environmental standpoint.

In the same way, companies A and B advertise their efforts for improving employees' satisfaction and social responsibility towards them. There is a slight differences between companies A and B. The managers who belong to company A, which sells standard products, underlined that it is not that easy to advertise CSR initiatives through the product. They have typically to reach their customers through specialised magazines or newsletter; while company B, more focused on innovative products can also use sponsorship to technical conferences or research. For both A and B, senior managers confirmed that their customers, usually industries in rich markets, take for granted internal investments for employees' satisfaction and they appreciate a less aggressive marketing on CSR initiatives based on those particular cultural channels.

Anyhow, all the four companies choose carefully their dealers. Dealers are considered the direct interface to customers and therefore they have to respect CSR principles as well.

5.4 Budgeting Marketing Plans and Actions

When the four companies come to the budget for their marketing plans and actions it is interesting to evaluate whether or not there are differences in terms of money spent on them. Starting with companies C and D which sell to poor markets, the senior managers reported that CSR marketing plans can absorb on average 10 % of all marketing initiatives. Companies A and B, which sell to rich markets, were a little bit more precise showing some figures. For instance company A in the last three years spent on average 11.2 % of its entire marketing budget and company B 10.7 %. Considering that the four companies have similar turnovers it can be assumed that there are not significant differences.

5.5 Reviewing Objectives, Corrective Actions

Marketing objectives are usually reviewed in all the four companies in a typical monthly management review meeting. All the marketing objectives, including CSR marketing objectives, are analysed in their performances trying to evaluate if they meet the targets. It is interesting to notice how all the four companies have adopted a flexible approach to the reviewing process. As a senior manager reported, CSR for its nature is particularly linked to laws, regulations and new industry standards, often different with each other depending on the country or market. Therefore, it can happen that even though you have set your CSR objectives and targets, you have to quickly react changing them. This reviewing and changing speed has become quicker in the last decade due to ethical scandals and the consequent introduction of new regulations.

5.6 Shaping the Theoretical Framework

From the theoretical results above analysed and discussed, a theoretical framework can be shaped. With the due limitations related to this particular qualitative case study, quadrant in Fig. 3 can be completed with the characteristics of the four companies.

From the model it can be learnt that there is a first fundamental difference between companies which sell in poor markets and companies which sell in richer ones. The formers are more linked to external CSR marketing strategy such as

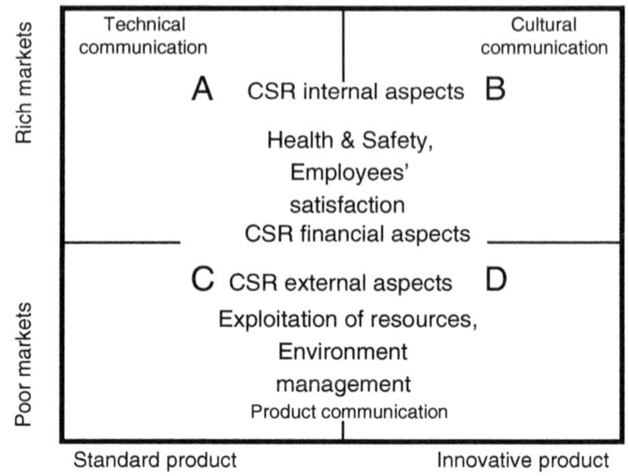

Fig. 5 The theoretical framework

exploitation of resources and environment management, while the latters are more linked to internal marketing strategy such as health and safety and employees' satisfaction. For all the company it is fundamental to define and respect ethical strategies connected to CSR financial aspects.

There are differences concerning how the companies communicate their CSR strategies and initiatives to customers and markets. For companies which sell in poor markets it seems that a communication based on the product and its characteristics could be more suitable. While for richer markets companies prefer a more technical and cultural communication based on specialised magazine, conferences and research. Anyhow, communication is the most important CSR marketing leverage.

6 Conclusions

Literature review has demonstrated how message content and communication channels are specific factors which influence the effectiveness of CSR communication. Furthermore some authors (McWilliams and Siegel 2001) have drawn the attention to segmentation of markets as an important aspect for satisfying the different stakeholders from a CSR point of view.

The theoretical framework which comes out from this research and is shown in Fig. 5 agrees with these results. The four companies of the case study are intensely using several communication channels even if they differ each other depending on the kind of market. Anyhow, communication as well as segmentation is a fundamental aspect of the MS dedicated to CSR.

As a novelty from this research comes up that the budget and efforts for advertising CSR strategies and initiatives are similar for all the markets and products. Just the communication ways are different. It seems that CSR can pay off companies which sell both innovative and less innovative products. This is in disagreement with some literature review results (Bhattacharya and Sen 2004) which consider CSR actions not that suitable for less innovative companies. Moreover, all the companies, including companies with less innovative products, invest in CSR initiatives in order to avoid financial scandals.

Another novelty could be related to the reviewing process of the strategic CSR objectives. The four companies demonstrated that, nowadays, this process can lead to quick changes in the CSR strategies whereas in the past was more stable.

This research presents some limitations, mainly due to its exploratory nature using case studies. It is difficult to generalise theoretically results from a qualitative inquiry like this. Anyhow, this research offers new avenues of research; first of all scholars should better investigate these results using a quantitative inquiry and a large sample of companies. Besides, this research is limited only to European large manufacturing companies; thence the investigation should be enlarged to small and medium sized companies as well as service industry.

References

Andrews KR (1997) The concept of corporate strategy. In: Foss NJ (ed) Resources, firms and strategies: a reader in the resource-based perspective. Oxford University Press, Oxford

Baccarani C (2008) What does ethical behaviour mean in management activities? The TQM J 20 (2):154–165

Bhattacharya CB, Sen S (2004) Doing better at doing good. Calif Manage Rev 47(1):9–24

Bhattacharya CB, Smith NC, Vogel DJ (2004) Integrating social responsibility and marketing strategy: an introduction. Calif Manage Rev 47(1):6–8

Brown TJ (1998) Corporate associations in marketing: antecedents and consequences. Corp Reputation Rev 1(3):215–233

Brown TJ, Dacin PA (1997) The company and the product: corporate associations and consumer product responses. J Mark 61(January):68–84

Burke L, Logsdon JM (1996) How corporate social responsibility pays off. Long Range Plan 29 (4):495–502

Campbell DT (1979) Qualitative and quantitative methods in evaluation research. Sage, London

Cravens D, Piercy NF (2008) Strategic marketing. McGraw-Hill Irwin, New York

Creswell JW (1998) Qualitative inquiry and research design: choosing among five traditions. Sage, London

Denzin NK (2000) Handbook of qualitative research. Sage, London

Dogan M, Pelassy D (1990) How to compare nations: strategies in comparative politics. CQ Press, Washington DC

Du S, Bhattacharya CB, Sen S (2010) Maximizing business returns to corporate social responsibility (CSR): the role of CSR communication. Int J Management Rev 12(1):8–19

Freeman RE (2001) Stakeholder theory of the modern corporation. In: Hoffman WM, Frederick RE, Schwartz MS (eds) Business ethics: readings and cases in corporate morality, 4th edn. McGraw Hill, Boston

Hillestad SG, Berkowitz EN (2012) Health care marketing plans: from strategy to action. Jones and Bartlett Learning, Burlington

Jocumsen G (2004) How do small business managers make strategic marketing decisions?: a model of process. Eur J Mark 38(5/6):659–674

Klein J, Dawar N (2004) Corporate social responsibility and consumers' attributions and brand evaluations in a product–harm crisis. Int J Res Mark 21(3):203–217

Kotler P (1989) From mass marketing to mass customization. Strategy Leadersh 17(5):10–47
Lantos GP (2001) The boundaries of strategic corporate social responsibility. J Consum Mark 18 (7):595–632
Lofland J, Lofland LH (1995) Analyzing social settings: a guide to qualitative observation and analysis. Wadsworth, Belmont
Lord BW, Velez R (2013) Converge: transforming business at the intersection of marketing and technology. Wiley, New York
Lymbersky C (2008) Market entry strategies: text, cases and readings in market entry management. Management Laboratory Press, Hambourg
McDaniel SW, Rylander DH (1993) Strategic green marketing. J Consum Mark 10(3):4–10
McWilliams A, Siegel D (2001) Corporate social responsibility: a theory of the firm perspective. Acad Manage Rev 26(1):117–127
Mendleson N, Polonsky MJ (1995) Using strategic alliances to develop credible green marketing. J Consum Mark 12(2):4–18
Merriam SB (1988) Case study research in education: a qualitative approach. Jossey-Bass Publishers, New York
Mintzberg H (1994) Rise and fall of strategic planning. Simon and Schuster, New York
Nan X, Heo K (2007) Consumer responses to corporate social responsibility (CSR) initiatives: examining the role of brand-cause fit in cause-related marketing. J Advertising 36(2):63–74
O'Reilly K (2005) Ethnographic methods. Routledge, London
Pearsall M (1995) Participant observation as role and method in behavioral research. Nurs Res 14 (1):37–42
Piercy NF, Lane N (2009) Corporate social responsibility: impacts on strategic marketing and customer value. Mark Rev 9(4):335–360
Polonsky MJ (1994) An introduction to green marketing. Electron Green J 1(2):1–10
Ronald Q (2012) SWOT analysis: raising capacity of your organization. Rutgers school of social work, http://socialwork.rutgers.edu/Libraries/Huamin/Brochure_2.sflb. Accessed 05 Jun 2014
Sen S, Bhattacharya CB (2001) Does doing good always lead to doing better? Consumer reactions to corporate social responsibility. J Mark Res 38(2):225–243
Stake R (1995) The art of case research. Sage, London
Straughan RD, Roberts JA (1999) Environmental segmentation alternatives: a look at green consumer behavior in the new millennium. J Consum Mark 16(6):558–575
Wilson RM, Gilligan C (2005) Strategic marketing management: planning, implementation and control. Routledge, London
Yin RK (1994) Case study research: design and methods, 2nd edn. Sage, Beverly Hills

Sustainable Knowledge Management and the Outsourcing of Core Competences—Does that Fit Together? Initial Insights from a Literature Review

Susanne Durst, Ingi Runar Edvardsson and Guido Bruns

Abstract The issue of sustainability has been frequently discussed in recent years. Knowledge and its management represent aspects that fit nicely with one of the underlying conditions of sustainability, i.e. durability. Indeed, proper knowledge management can help firms in coping with present and future business challenges. The outsourcing of secondary business functions has become popular with many firms. What is increasingly present is the outsourcing of knowledge competences that is the firms' core competences. This trend means a challenge for both the firms' knowledge management and their sustainability. Against this background, the purpose of this paper is to review extant research on the outsourcing of knowledge competences to establish our body of knowledge and to identify gaps justifying further research activities. The study is based on a systematic review of peer reviewed empirical articles on the outsourcing of knowledge competences. This proceeding proves evidence that there are a small number of papers addressing outsourcing and its consequences for knowledge management (KM) sustainability. The recommendations derived from the findings can assist researchers, managers and consultants to better understand the link between knowledge competence outsourcing and KM sustainability. This increased understanding can particularly be useful for managers as better decisions will be possible.

Keywords Outsourcing · Knowledge management · Knowledge competences · Systematic review

S. Durst (✉)
School of Business, University of Skövde, Högskolevägen, 54128 Skövde, Sweden
e-mail: susanne.durst@his.se

I.R. Edvardsson
School of Business, University of Iceland, Gimli, 101, Reykjavik, Iceland
e-mail: ingire@hi.is

G. Bruns
Fagelas Bygg, Landbyn 1, 544 94 Hjo, Sweden
e-mail: gbbygg@yahoo.se

A. Chiarini (ed.), *Sustainable Operations Management*,
Measuring Operations Performance, DOI 10.1007/978-3-319-14002-5_2

1 Introduction

The pursuit of sustainable business development has been subject to an increasing number of organizations. In contrast to the past, business activities of organizations reveal a stronger emphasis on societal and environmental issues compared to economic issues which dominated in many organizations. Research on sustainable business development has also increased, yet it is still in its infancy (Iasevoli and Massi 2012). Sustainable development is defined as "development that meets the needs of the present without compromising the ability of future generations to meet their own needs" (WCED 1987: 43). In order to do so firms are supposed to adopt social, economic, and environmental development in their business operations (Chow and Chen 2012). Thereby these three aspects need to be considered as one entity rather than as separate entities (Van Kleef and Roome 2007). Bansal (2005) explained sustainable development using the resource-based view (RBV) and the institutional based view. The former is primarily interested in developing and exploiting resources and capabilities for generating rents (Roos 2005). Especially intangible assets or knowledge resources are associated with the attainment of this objective (Carmeli 2004), as these resources are more likely of being scarce, durable, valuable, inimitable, and non-substitutable than tangible assets (Barney 1991; Grant 1991). Investments in sustainable practices are expected to increase both an organization's competitive advantage and its operational performance (Schoenherr 2012; Iasevoli and Massi 2012). Thereby, not surprisingly, performance is not only considered in terms of economic performance (for example savings or profitability) but also in non-economic terms such as eco-efficiency (Iasevoli and Massi 2012). In order to reach the sustainable goals set, collaboration with all stakeholders is considered essential (Van Kleef and Roome 2007).

Against this background knowledge and knowledge management (KM) can be regarded as central to sustainable business development and continuity of organizations (Gloet 2006; Robinson et al. 2006). KM is "a set of procedures, infrastructures, technical and managerial tools, designed towards creating, circulating (sharing) and leveraging information and knowledge within and around organisations" (Bounfour 2003: 156). This definition clarifies that KM has a long-term orientation and therefore fits nicely with one of the underlying assumptions of sustainability and sustainable management, which is durability (Chow and Chen 2012). KM practices (such as knowledge creation, knowledge dissemination, knowledge storage and knowledge application), are expected to support a continued development of knowledge that is up to date and relevant. Consequently, these KM practices make sure that organizations are able to address current and future business challenges (Durst and Edvardsson 2012), which in turn help them in meeting their aims concerning sustainable business development (Gloet 2006). Hence, sustainable knowledge management can be regarded as a synopsis of KM and sustainability; where the primary role of KM is to treat current and future knowledge resources sustainably by considering social, economic, and environmental aspects (Ansari et al. 2010).

Outsourcing has become rather common in recent years and organizations pursue different objectives with this approach. Perhaps the most prominent reason of organizations for outsourcing is to better focus on their core competences (Wu et al. 2005). Another widely cited rationale is to gain access to unique resources, skills and talents, and capabilities possessed by other firms (Howells et al. 2008). Greater flexibility in managing demand swings and reducing company risk by sharing it with suppliers are mentioned as other motives (for example Kremic et al. 2006; Di Gregorio et al. 2009; Quinn 1999). In sum, one can argue that all these reasons for outsourcing are expected organizations on their search for competitive advantage (Mudambi and Tallman 2010). Despite the possible benefits, there are also some indications that outsourcing can undermine the knowledge base of companies, which means that outsourcing might only be beneficial in the short run (Xia and Tang 2011). Schlosser et al. (2006), for instance, argue that outsourcing weakens organizational learning as its temporary nature undermines the traditions and routines associated with a strong organizational culture. Similarly, firms that outsource their core business functions may lose touch with their technological know-how that provides opportunities for product and process innovation (Sánchez et al. 2007). Consequently, if the negative aspects of outsourcing overtake the positive ones, the knowledge (management) of the firms concerned is at risk, that is its sustainability is threatened. It can also have serious implications for the firms' overall sustainable development. Still one can observe that some organizations (for example pharmaceutical organizations) have started outsourcing parts of their key business functions (Howells et al. 2008) which may be contradictory to both sustainable KM and sustainable development. What prompt organizations to outsource their knowledge competences as well?

Having this in mind, the purpose of this paper is to review extant research on the outsourcing of knowledge competences to establish our body of knowledge. Thereby, in this paper outsourcing is defined as the allocation or reallocation of an organization's internal activities and services to outside providers. These providers can operate in the same country (onshore outsourcing) or in another country (offshore outsourcing). Knowledge competences, on the other hand, refer to a specific set of abilities that enable organizations to compete and develop in order to meet current and future business requirements. To the authors knowledge this paper is unique in three ways: (1) Research has so far neglected the study of knowledge competence outsourcing, (2) no research has discussed knowledge competence outsourcing from a sustainable KM point of view, and (3) the intersection of sustainability and outsourcing in general represents an under researched field of study (Babin and Nicholson 2011).

The remaining paper is organised as follows. In the next section the related literature is briefly outlined. Next the method employed is briefly discussed. This is followed by a presentation of the results. The paper terminates with the conclusion and implications of the study.

2 Literature Background

Much of the existing literature on KM is concerned with how organisations can capture knowledge from experts within the organisation, and formalise and package knowledge assets for dissemination and reuse by other employees (Markus 2001). As over the years the business environment has become more complex organizations have been forced to expand knowledge capture to external individuals/organizations as well in order to have access to a broader knowledge base (Majchrzak et al. 2004). Various mechanisms and structures, often technology-supported, have been proposed to facilitate this endeavour including Intranets (McKinlay 2002), discussion forums (Hansen and von Oetinger 2001), lessons learned databases (Brown and Duguid 2000), expert yellow pages (Storey and Barnett 2000), mentor groups (Zack 1999) and communities of practice (Wenger et al. 2002). The emphasis has also been on transferring tacit knowledge into explicit (Nonaka 1991). Even though certain KM practices, such as knowledge creation, have been expanded to include external parties as well, when it comes to the core of their business activities organizations usually prefer to build upon an internal KM strategy where the focus is on leveraging existing knowledge from within the boundaries of the organisation. This paper addresses an alternative and less widely discussed view of KM, referred to here as "knowledge outsourcing" (KO). In KO, external experts are explicitly contracted to generate knowledge-intensive assets, which are subsequently internalised by the organisation. Hence, KO exploits external sources for knowledge creation rather than relying on internal capabilities. KO might be interesting for organizations that have already developed a comprehensive knowledge about outsourcing and thus are willing to go a step further (Babin and Nicholson 2011). While the general concept of outsourcing is not new, as evidenced by the wealth of published literature in the area of information systems (for example McKeen et al. 2002; Lam and Chua 2009) and more detailed in the following, the study of KO, however, is in its infancy (Tarn and Chien-Chih 2012).

Along with an increased importance of services and knowledge in the economy, in a number of firms organizational restructuring has been taken place, such as intensified networking and changes in the supply chain manifested in outsourcing. Outsourcing has grown rapidly among public and private organizations in recent years (see Bryson 2007; Di Gregorio et al. 2009; Kakabadse and Kakabadse 2002). In all cases, outsourcing always requires a third party involvement (Jagersma and van Gorp 2007).

Outsourcing is a strategic move which involves both sourcing absent activities that new firms may not have completed in-house in the past, or the substitution of internal activities by transferring these, in part or whole, to a third party supplier that performs the task, function, or process (Gilley and Rasheed 2000; Holcomb and Hitt 2007). Advances in information and communication technologies have enabled new firms to pursue the outsourcing of value-creating activities such as software development, engineering, and research and development (Hui et al. 2008). It has also contributed to the development of a new business industry in countries, such as software outsourcing in India (Jain and Khurana 2013). Outsourcing projects are highly dependent on knowledge sharing in order to make them work, thus proper KM systems are crucial (Aydin et al. 2010) which highlights the link between outsourcing and KM.

To date, researchers have focused on outsourcing by large, established firms (Bhalla et al. 2008; McIvor 2009); however, there is evidence that outsourcing attracts new firms as well. Mills (2002), for example, showed in his study how new biotechnology firms utilize intermediate markets for a variety of value chain activities. Why might new firms outsource activities, including value-creating activities such as research and development, which are known to contribute to the value-creating potential of firms (Kumar et al. 2009)? Researchers who stress the integrated view of transaction cost theory (TCT) and RBV argue that by establishing relationships specifically with well-known firms, new firms can not only reduce the search and monitoring costs associated with finding a reliable partner but also acquire recognition and use their established counterparts to draw vital combinations of resources such as status and physical resources (Lin et al. 2009). This is crucial for new firms as they suffer from restrictions such as scarcity of human capital and operational know-how as a consequence of liabilities of newness and smallness (Aldrich and Auster 1986; Baum and Oliver 1992; Stinchcombe 1965; Bhalla and Terjesen 2013). Regardless of the age of companies, one may ask why organizations would outsource critical knowledge competences as it may reduce the chance of serving their customers and other stakeholders successfully or of impacting their long-term success (Howells et al. 2008). Indeed knowledge competences are viewed as the critical resources needed to develop and survive (Parry et al. 2006) and to deliver value added services and products (Wu et al. 2005). Is it mainly the hope of short-term gains or is KO the outcome of a well-considered decision? Given the competitive fierce many organizations are exposed to one may assume the former may dominate. Additionally, many firms may not be able to make a clear distinction between core and non-core activities, so as a consequence they unintendedly outsource core competences.

Outsourcing business activities is mainly found among US and UK organizations, whereas Italian, French and German firms appear to be more reluctant concerning this approach (Parry et al. 2006). This may indicate that the former are more short-term oriented (i.e. seeing outsourcing as a means of achieving competitive advantage) then the latter ones. Or as Xia and Tang (2011: 504) put it "the current low-cost purchasing/outsourcing strategy comes from the voice to focus on "here and now" and on survival." This observation also suggests that US and UK firms have fewer problems with giving up control. As sustainable development is concerned, it would require an increasing focus on social responsibility concerning outsourcing in general (Gloet 2006).

To sum up, outsourcing represents a topic that has frequently been studied in general. Thereby the focus has been on studying outsourcing activities of large and established companies. Recent developments suggest that outsourcing has started going beyond secondary (non-core) business functions, but covering core knowledge competences as well. Competences that form the basis of competitive advantage, thus addressing an organization's core business function(s). This situation may put the organization at risk, particularly from a viewpoint of sustainable KM and sustainable business development.

3 Methodology of Literature Review

In our review process, we adopted the principles of a systematic review as recommended by Jesson et al. (2011) namely:

1. Mapping the field through a scoping review—This means that one establishes the known and any knowledge gaps.
2. Comprehensive search—This means that one reverts to the different electronic databases available and start searching within those databases using a number of predefined keywords as well as other inclusion criteria (for example peer reviewed articles, English language).
3. Quality assessment—This comprises the reading (that is scanning the abstract and if necessary more sections of the paper) of the papers identified and the decision about whether or not to include them in the review.
4. Data extraction—This refers to the collection of relevant data from the selected papers and the capturing of the data into a pre-designed extraction sheet.
5. Synthesis—This comprises the synthesis of the data from each individual paper into one sheet in order to show the known and to provide the basis for establishing the unknown.
6. Write-up—This final stage is about the production of the report about the review.

In the following we explain how we dealt with the six steps. First, we developed a research plan comprising the research questions we were interested in answering. This also involved the keywords, and a set of inclusion and exclusion criteria. We were interested in the current status of research on outsourcing of knowledge processes in order to identify promising areas for future research. The questions formulated, as outlined above, were: (1) Which topics are well researched in relation to outsourcing and which are not? (2) Which were the main findings of the studies?

We decided to use multiple keywords to identify relevant studies, such as outsourcing, knowledge competence, knowledge-based service. Our inclusion criteria were: empirical research papers, peer reviewed, English language, ProQuest and Web of Science databases. We excluded grey literature such as reports and non-academic research, other languages than English, and other databases than ProQuest and Web of Science.

Additionally, we produced an excel data sheet consisting of key aspects related to our research aim. In our case these were: name of author(s), year of publication, research aim/objectives, theoretical perspective/framework, method, main findings, and name of the journal.

Second, once we had specified all the relevant issues, one of us accessed ProQuest and Web of Science and searched using combinations of the keywords set. We looked for combinations of these keywords in the title, keywords and abstract. The literature review included papers published until 2 July, 2014. Depending on the combination of keywords used, different numbers of hits were generated.

Third, each of us manually scanned the abstracts of the respective papers and, if relevant, more parts of the articles to make sure that they actually fell within our scope of interest. This reduced the number of articles without duplications to the final number of 13 articles which fulfilled our criteria and were then analysed.

Fourth, all authors read the 13 papers individually and entered relevant data regarding our research purpose in the excel sheet.

Fifth, in the next stage we discussed the findings, synthesised the individual data into one and identified themes. This helped us to establish the current body of knowledge with regard to outsourcing (and offshoring) of knowledge processes.

Sixth, the final stage of our review process was devoted to the write-up of our findings.

4 Presentation of Findings

Among the thirteen papers that formed the basis for our analysis, the oldest publication is from 2006 and the most recent ones are from 2013 (two papers). One paper was published in 2008, four of the papers in 2009, four papers in 2010, and one in 2011. No papers were published in 2007 and 2012.

In the sections below we present our analysis concerning the following aspects: general observations which outline the research methods applied, the research setting chosen and the journals in which the papers were published. After that, the study's main findings according to the themes identified are presented.

4.1 General Observations

With regard to the methodology, the most common method applied is the case study approach (eight papers). Additionally, surveys (three papers), focus groups, and secondary data (one each) were used as research methods. We identified two papers which was longitudinal in nature (Brege et al. 2010; Cheng et al. 2010).

As the national research setting is concerned, the most investigated research setting is Europe (6 papers). Four studies were found in Asia, one in the US, and one paper were based on an international setting (dominated by the US). One paper could not be assigned to any national setting.

The 13 papers were published in 11 different journals. Six of the 13 journals can be assigned to the field of operations, technology and management, two journals to international business, and the remaining journals address fields such as innovation, marketing, entrepreneurship and small business management, and organization studies. This suggests that the topic appeals to a broad readership. It may also suggest that the topic has not found its strategic home yet.

4.2 Body of Knowledge Regarding the Outsourcing of Knowledge Competences

We summarized the main findings of the reviewed paper under five broad themes:

- Reasons for and against KC outsourcing;
- Consequences of knowledge competence (KC) outsourcing;
- Contributing factors to the success of KC outsourcing;
- Framework/Theory development;
- Others.

The following sections provide insights into the body of knowledge established concerning the five different themes.

4.2.1 Reasons for and Against KC Outsourcing

Understanding the reasons and motives why firms outsource specific business functions has been the focus of four papers (Table 1). Two papers mainly addressed reasons for KC outsourcing. For example, having studied outsourcing over an extended period, Brege et al. (2010) study highlighted besides lower costs, development of core competences, control and flexibility, the potential of outsourcing for assisting in business development. Whereas, Howells et al. (2008) study listed main reasons for R&D outsourcing. These are access to expertise which is not available in the firm, reduction of development time, time for product launch and costs, and support for technology change. Lau and Zhang (2006), who studied both reasons for and against KC outsourcing, concluded that the reasons for outsourcing can be found in three areas: economic, strategic and environmental. In the same vein, Rundquist and Halila (2010) determined that the main reason for KC outsourcing, as found in successful firms, is the need for knowledge that is missing in the firm.

The last two papers also presented insights into reasons against KC outsourcing. Lau and Zhang (2006), for instance, considered the following as hampering factors: Lack of capable service providers, loss of control, poor transportation and IT infrastructure, local protection regulations, and lack of overall post-outsourcing measurement. Rundquist and Halila (2010), on the other hand, stressed strategic decisions from the board and risk of competence drainage as main reasons against KC outsourcing.

Table 1 Papers on reasons for and against KC outsourcing

Author(s)	Year	Research aims/objectives	Theoretical perspective/ framework	Method (empirical/ theoretical)	**Main findings**	**Journal**
Brege et al.	2010	To describe and analyze how and why the question of outsourcing or insourcing within a specific empirical context, telecommunications services, has been handled differently over time	Two theoretical approaches: transaction cost analysis and core competence thinking	Longitudinal case study on procurement of telecommunications for the Swedish public sector's organizations during the last 40 years	In addition to low cost, core competence, control and flexibility, the paper identifies business development as an additional logic for outsourcing and insourcing	Strategic Outsourcing: An International Journal
Lau and Zhang	2006	To explore the key factors that motivate organizations in China to outsource and the obstacles these companies are facing in comparison with the situation in Western developed countries	Literature on the drivers, obstacles and problems of outsourcing	A case study approach involving six Chinese companies of three major ownership types in China (i.e. state-, private- and foreign-owned)	Economic, strategic, and environmental factors are the main drivers to engage in outsourcing. Lack of capable service providers, loss of control, poor transportation and IT infrastructure, local protection regulations, and lack of overall post-outsourcing measurement are considered the main obstacles and problems in the outsourcing process. The findings are summarized in a framework for making outsourcing decisions in China	International Journal of Physical Distribution and Logistics Management
Rundquist and Halila	2010	To improve our understanding of new product	Discussion of NPD outsourcing	E-mail survey sent to Swedish medium-sized	Most important reason for NPD outsourcing is the	European Journal of

(continued)

Table 1 (continued)

Author(s)	Year	Research aims/objectives	Theoretical perspective/ framework	Method (empirical/ theoretical)	**Main findings**	**Journal**
		development (NPD) outsourcing. The authors are specifically interested in factors affecting the outsourcing decisions	from a resource-based perspective and a transaction cost perspective	firms in four industries (Mechanical, Food, Plastic and Rubber, Wood and Paper)	need for knowledge that does not exist in the firm. Reasons to not outsource activities are: strategic decisions from the board and risk of competence drainage. Personal interaction is an important factor for the integration of process knowledge, this was found with the best firms group	Innovation Management
Howells et al.	2008	To outline the main reasons for pharmaceutical firms to outsource R&D	n/a	Survey of pharmaceutical enterprises in the UK, 105 responses were received	The authors present a typology of outsourcing factors in relation to R&D. The findings suggest that the outsourcing activities are mainly found with regard to applied activities, i.e. activities that are less core to the organizations, e.g. clinical trial activity, R&D software and applied research	R&D Management

4.2.2 Consequences of KC Outsourcing

Two papers were assigned to this topic (Table 2). Agndal and Nordin (2009) highlighted in their paper negative (non-financial) consequences KC outsourcing can mean to firms (i.e. loss of knowledge and skills). Based on their findings the authors propose ways to overcome them. In this context, the authors stressed in particular the relevance of being aware of possible consequences as a critical starting point for addressing them. Bhalla and Terjesen (2013) showed how KC outsourcing can help firms in expanding their supplier network and operational knowledge base. However, to actually benefit from it firms need to develop their technical, evaluation, relational, entrepreneurial, and integration competencies. Additionally, the authors' findings point at the learning opportunity those supplier networks can trigger. Finally, joint activities with supplier networks can contribute to reduced relationship specific investments and cost efficiency.

4.2.3 Contributing Factors to the Success of KC Outsourcing

Two papers were assigned to this theme (Table 3). Bustinza et al. (2010) demonstrated the contribution of learning capacities of firms and their stock of knowledge resources to the success of outsourcing projects. The authors' study showed that both can positively influence the firms' search for competitive advantage and the way as to how the relationship with outsourcing partners is managed. On the other hand, by looking at KC outsourcing from a risk perspective, Mahmoodzadeh et al. (2009) work, that examined outsourcing in the context of business process management and knowledge management, identified the following issues possible hampering the success of outsourcing projects: "(1) dependency on the suppliers and changing collaborative to opportunistic behavior of the supplier; (2) losing touch with new technological opportunities for product and process innovations; (3) communication and coordination problems; (4) cognitive distance between suppliers and firm therefore makes it more difficult to align decisions and exchange knowledge; (5) outsourcing functions that should not be outsourced; and (6) decreasing control over the outsourced functions" (861).

4.2.4 Framework/Theory Development

Three papers can be assigned to this topic (Table 4). These papers aim at providing theoretical frameworks and development of existing theory in order to progress and expand our understanding of knowledge outsourcing. Chang et al. (2009) propose an integrated model consisting of three different theoretical perspectives (i.e. technological regime, resource-based view and transaction cost theory) as a basis for the identification of factors influencing outsourcing strategies. By having such an integrated approach the authors believe that better decisions regarding suitable outsourcing strategies are possible. Cheng et al. (2010), on the other hand,

Table 2 Papers on the consequences of KC outsourcing

Author(s)	Year	Research aims/objectives	Theoretical perspective/ framework	Method (empirical/theoretical)	Main findings	Journal
Agndal and Nordin	2009	The purpose of this paper is to develop a model that enables the study of non-financial consequences of outsourcing	Literature on outsourcing of different company functions	Literature review and focus group involving experienced senior procurement managers	The study highlights different negative non-financial consequence of outsourcing (e.g. loss of knowledge and skills etc.). In addition many examples are provided as to how the participants tried to overcome potentially negative consequences through good business practices or where such consequences were perceived as irrelevant due to the outsourcing context	Benchmarking: An International Journal
Bhalla and Terjesen	2013	To analyse outsourcing by knowledge-intensive new firms in supplier networks	Transaction cost theory and resource based view	Case study based on data from ten biotechnology start-ups and twenty of their suppliers	The research reveals that new firms outsourcing to highly-embedded suppliers are likely to secure access to a wider supplier network, attain best-in-class operational knowledge, and avoid supplier opportunism while facing low levels of relationship-specific investments. New firms outsourcing to suppliers at the network periphery are more likely to realize cost efficiencies, expose themselves to opportunism, uncertainty, and higher levels of relationship-specific investments but low levels of operational knowledge. The authors propose five outsourcing competencies new firms should develop to realize the benefits	Industrial Marketing Management

Table 3 Papers on contributing factors to the success of KC outsourcing

Author(s)	Year	Research aims/objectives	Theoretical perspective/ framework	Method (empirical/theoretical)	Main findings	Journal
Bustinza et al.	2010	To analyse outsourcing from the point of view of knowledge management and learning	Knowledge-based theory of the firm	A survey among 204 Spanish service firms with more than 20 employees. Reached a response rate of 20.4 %	Findings highlight the contribution of the stock of knowledge as a resource and of learning capacity to both competitive advantage and the capacity to manage outsourcing collaborations	Journal of Supply Chain Management
Mahmoodzadeh et al.	2009	To analyse the impact of business process management (BPM) and knowledge management (KM) on reduction of outsourcing risks and pitfalls	Literature on risks of outsourcing and outsourcing models and frameworks	A literature review was accompanied by a case study conducted on MTN Irancell	The authors propose a comprehensive framework to help performing each step of the business process outsourcing (BPO) lifecycle and the reduction of BPO risks and pitfalls using BPM and KM approaches	Business Process Management Journal

Table 4 Papers on framework and theory development

Author(s)	Year	Research aims/objectives	Theoretical perspective/framework	Method (empirical/theoretical)	Main findings	Journal
Chang et al.	2009	To examine the strategic technology outsourcing of corporate ventures from an integrated perspective	Review of factors influencing technology sourcing from three theoretical perspectives: technological regime, resource-based view, and transaction cost theory	Case study approach, study of four Taiwanese publicly traded pharmaceutical companies	The authors propose an integrated model that suggests that the factors of technological regimes, resource-based view, and transaction cost perspective together affect a firm's sourcing strategy	International Journal of Entrepreneurial Behaviour and Research
Cheng et al.	2010	To explore how to transfer production know-how on the shop-floor level when manufacturing units are relocated.	Literature on knowledge transfer	Case study method involving three large Danish manufacturing companies. Longitudinal (2 years)	The authors introduce different means for knowledge transfer and classify them according to their usages. Additionally a means framework is proposed to integrate all the elements systematically	Strategic Outsourcing: An International Journal

(continued)

Table 4 (continued)

Author(s)	Year	Research aims/objectives	Theoretical perspective/ framework	Method (empirical/theoretical)	Main findings	Journal
Lam and Chua	2009	The purpose of the paper is to present the notion of knowledge outsourcing as an alternative strategy of knowledge management	Literature on KM and IT outsourcing	A case study based on Fenton University (higher education enterprise owned by a consortium of international, research intensive universities)	The authors propose a general process model of knowledge outsourcing. They draw attention to three conditions under which knowledge outsourcing may be a suitable strategy for knowledge management. Additionally, two main areas of knowledge outsourcing risk are identified	Journal of knowledge management

developed an integrated framework based on a classification of different means for knowledge transfer according to their usages (e.g. through prototypes, documents or peer-to-peer training). Lam and Chua (2009), on the other hand, regard knowledge outsourcing as a promising alternative strategy of KM, thus pointing to a further development in the study of KM. The authors specify certain conditions which support decision making in favour of knowledge outsourcing. These are

> "lack or unavailability of in-house expertise; availability of external knowledge providers who are able to satisfy an organisation's knowledge needs; and a favourable business case in which the cost-benefit of knowledge outsourcing is positive in light of other available alternative options" (39).

4.2.5 Others

Two papers were assigned to this topic (Table 5). These papers have addressed the management of outsourcing and the application of outsourcing in a specific industry. Hamzah et al. (2011) who mainly studied knowledge management practices as found in an offshoring outsourcing company. Lacity and Willcocks (2013) studied Legal Process Outsourcing and mainly demonstrated the opportunities available in this new field of KC outsourcing.

When bringing together these results with the ones from the literature review KC outsourcing can be determined as a further strategy firms are pursuing in order to search for competitive advantage. The findings also show that the main focus is on KC outsourcing in large companies. As reasons against KC outsourcing are concerned, the comparison shows that they are basically in line with general reasons against outsourcing. Concerning KC outsourcing, however, a stronger emphasis on

Table 5 Papers on other KC outsourcing issues

Author(s)	Year	Research aims/objectives	Theoretical perspective/ framework	Method (empirical/ theoretical)	Main findings	Journal
Hamzah et al.	2011	To investigate how offshoring outsourcing company manage its knowledge resources	Literature on KM and offshore outsourcing	Kind of case study approach involving interviews and observations in an offshoring outsourcing support centre in Malaysia	The findings mainly highlight KM practices in the case company	International Journal of Business and Social Science
Lacity and Willcocks	2013	To assess the current legal process outsourcing (LPO) provider landscape	N/A	LPO provider data collected in 2011 by Orbys, a European sourcing and transformation advisory firm	The LPO providers provide a variety of services, including litigation, intellectual property, corporate, compliance, procurement, employment, property, and consulting services	Strategic Outsourcing: An International Journal

the long-term consequences can be established, which are addressing a firm's future capability of keeping and developing its competitive advantage. Despite this the findings from the systematic literature review suggest that KC outsourcing has the potential of helping organizations to pursuit sustainable KM and sustainable business development as learning from the past and improved collaboration are possible, which can be ascribed to the long-term perspective of KC outsourcing as well.

5 Conclusions

The starting point of this paper was to bring together the issues of sustainable KM and the outsourcing of core competences. As one may argue that the latter presents a severe damage of the former. Given the fact that an increasing number of firms pursuit a sustainable development of their organizations, of which sustainable KM is a critical part, one may wonder why still some firms are outsourcing their core competences, that is those knowledge assets that are critical for company success and competitive advantage.

In order to address the situation outlined above, this paper reviewed extant research on outsourcing and knowledge competences to establish our understanding. We used the method of a systematic literature review to identify relevant papers. We found thirteen papers that formed the basis of our analysis. The papers were published in the years 2006–2013, highlighting the topicality of the topic, and the small number of papers indicates that it is an under researched field of study.

The main findings were summarised in five themes: Reasons for and against KC outsourcing; consequences of KC outsourcing; contributing factors to the success of KC outsourcing, framework/theory development, and other issues. Concerning the issue of sustainable business development as well as sustainable KM, the findings regarding the consequences of KC outsourcing and contributing factors to the success are of specific relevance as they are addressing a firm's long-term position and therefore underlying assumption of any activities following a sustainable approach. Indeed the findings suggest that KC outsourcing can be a strategy for reaching sustainable business development, even though the pros and cons of KC outsourcing are close together.

From a sustainable point of view, aspects in favour of KC outsourcing are having access to a stronger and broader network, developing the operational knowledge base, reduced relationship specific investments and learning. Indeed the latter can in combination with the knowledge stock at hand contribute to competitive advantage. On the other hand, aspects speaking against KC outsourcing are the loss of knowledge and skills, for example, the partial or fully loss of a firm's ability to perform certain business functions (Agndal and Nordin 2009). This can also be accompanied by a loss of control over the business activities undertaken meaning that a firm is at someone else's mercy.

The findings have further suggested that in order to make use of the benefits of KC outsourcing firms need to develop a new/different set of competencies and must be aware of and take responsibility for the consequences of their actions regarding KC outsourcing. Consequently, KC outsourcing is mainly something for firms who are able to follow a consistent and disciplined approach to KC outsourcing. Against the background of observable business practices that are characterized by permanent (and often radical) policy changes, this ability will only show in a very small number of organizations. Therefore, from a practical point of view, this study provides insights into the usefulness of KC outsourcing in conjunction with sustainable business development. The findings also offer guidance for managers about the preconditions that must be fulfilled in order to apply KC outsourcing as part of a sustainable KM strategy. In this context, it might be promising to examine whether public organizations are better suited for KC outsourcing compared to private organizations.

The present study is not without limitations. A complete coverage of all the articles considering the issue of knowledge competence outsourcing could not have been achieved, given the search proceeding chosen. So it may have left out papers that also addressed the topic but used a different language. Yet, it seems reasonable to assume that the review process covered a large proportion of the empirical studies available.

References

Agndal H, Nordin F (2009) Consequences of outsourcing for organizational capabilities: some experiences from best practice. Benchmarking Int J 16(3):316–334

Aldrich H, Auster E (1986) Even dwarfs started small: liabilities of age and size and their strategic implications. In: Staw B, Cummings LL (eds) Research in organizational behavior, vol 8. JAI Press, Greenwich, pp 165–198

Ansari CF, Holland A, Fathi M (2010) Advanced knowledge management concept for sustainable environmental integration. In: IEEE 9th international conference on cybernetic intelligent systems (CIS), 1–2 Sep 2010, pp 1–7

Aydin MN, de Groot J, van Hillegersberg J (2010) Action readiness and mindset for IT offshoring. J Enterp Inf Manage 23(3):326–349

Babin R, Nicholson B (2011) How green is my outsourcer? Measuring sustainability in global IT outsourcing. Strateg Outsourcing Int J 4(1):47–66

Bansal P (2005) Evolving Sustainably: a longitudinal study of corporate sustainable development. Strat Mgmt J 26(3):197–218

Barney JB (1991) Firm resources and sustained competitive advantage. J Manage 17(1):99–120

Baum J, Oliver C (1992) Institutional embeddedness and the dynamics of organizational populations. Am Sociol Rev 57(4):540–559

Bhalla A, Terjesen S (2013) Cannot make do without you: outsourcing by knowledge-intensive new firms in supplier networks. Ind Mark Manage 42(2):166–179

Bhalla A, Sodhi M, Son B (2008) Is more IT offshoring better?: an exploratory study of western companies offshoring to South East Asia. J Oper Manage 26(2):322–335

Bounfour A (2003) The management of intangibles: the organization's most valuable assets. Routledge, London

Brege S, Brehmer P-O, Lindskog H (2010) Sourcing, insourcing and two times outsourcing. Four phases of procurement of telecommunications services within the Swedish public sector. Strateg Outsourcing Int J 3(2):144–162
Brown JS, Duguid P (2000) The social life of information. Harvard Business School Press, Cambridge
Bryson JR (2007) The "second" global shift: the offshoring or global sourcing of corporate services and the rise of distanciated emotional labour. Geogr Ann Series B, Hum Geogr 89 (1):31–43
Bustinza OF, Molina LM, Gutierrez-Gutierrez LJ (2010) Outsourcing as seen from the perspective of knowledge management. J Supply Chain Manage 46(3):23–39
Carmeli A (2004) Assessing core intangible resources. Eur Manage J 22(1):110–122
Chang YC, Chen YC, Kuo TK (2009) Strategic technology sourcing in corporate ventures: a study of Taiwanese pharmaceutical firms. Int J Entrepreneurial Behav Res 15(5):497–517
Cheng Y, Madsen ES, Liangsiri J (2010) Transferring knowledge in the relocation of manufacturing units. Strateg Outsourcing Int J 3(1):5–19
Chow WS, Chen Y (2012) Corporate sustainable development: testing a new scale based on the mainland chinese context. J Bus Ethics 105(4):519–533
Di Gregorio D, Musteen M, Thomas DE (2009) Offshore outsourcing as a source of international competitiveness for SMEs. J Int Bus Stud 40(6):969–988
Durst S, Edvardsson IR (2012) Knowledge management in SMEs: a literature review. J Knowl Manage 16(6):879–903
Gilley K, Rasheed A (2000) Making more by doing less: an analysis of outsourcing and its effects on firm performance. J Manage 26(4):763–790
Gloet M (2006) Knowledge management and the links to HRM. developing leadership and management capabilities to support sustainability. Manage Res News 29(7):402–413
Grant RM (1991) The resource-based theory of competitive advantage: implications for strategy formulation. Calif Manage Rev 33(3):114–135
Hamzah N, Aman A, Maelah R, Amirudin R (2011) Managing knowledge resources in offshore outsourcing firm. Int J Bus Soc Sci 2(17):100–107
Hansen MT, von Oetinger B (2001) Introducing T-shaped managers. knowledge management's next generation. Harvard Bus Rev 79(3):106–116
Holcomb TR, Hitt MA (2007) Toward a model of strategic outsourcing. J Oper Manage 25 (2):464–481
Howells J, Gagliardi D, Malik K (2008) The growth and management of R&D outsourcing: evidence from UK pharmaceuticals. R&D Manage 38(2):205–219
Hui PP, Davis-Blake A, Broschak JP (2008) Managing interdependence: the effects of outsourcing structure on the performance of complex projects. Decis Sci 39(1):5–31
Iasevoli G, Massi M (2012) The relationship between sustainable business management and competitiveness: research trends and challenge. Int J Technol Manage 58(1/2):32–48
Jagersma PK, van Gorp DM (2007) Redefining the paradigm of global competition: offshoring of service firms. Bus Strateg Ser 8(1):35–42
Jain DM, Khurana R (2013) Need for sustainable global business model in software outsourcing: the Indian perspective. Bus Process Manage J 19(1):54–69
Jesson JK, Matheson L, Lacey FM (2011) Doing your literature review: traditional and systematic techniques. Sage, Los Angeles
Kakabadse A, Kakabadse N (2002) Trends in outsourcing: contrasting USA and Europe. Eur Manage J 20(2):189–198
Kremic T, Tukel IG, Rom OW (2006) Outsourcing decision support: a survey of benefits, risks, and decision factors. Supply Chain Manage Int J 11(6):467–482
Kumar K, Van Fenema PC, von Glinow MA (2009) Offshoring and the global distribution of work: implications for task interdependence theory and practice. J Int Bus Stud 40(4):642–667
Lacity MC, Willcocks LP (2013) Legal process outsourcing: the provider landscape. Strateg Outsourcing Int J 6(2):167–183

Lam W, Chua AYK (2009) Knowledge outsourcing: an alternative strategy for knowledge management. J Knowl Manage 13(3):28–43
Lau KH, Zhang J (2006) Drivers and obstacles of outsourcing practices in China. Int J Phys Distrib Logistics Manage 36(10):776–792
Lin Z, Yang H, Arya B (2009) Alliance partners and firm performance: resource complementarity and status association. Strateg Manag J 30(9):921–940
Mahmoodzadeh E, Jalalinia S, Nekui Yazdi F (2009) A business process outsourcing framework based on business process management and knowledge management. Bus Process Manage J 15 (6):845–864
Majchrzak A, Cooper LP, Neece OE (2004) Knowledge reuse for innovation. Manage Sci 50 (2):174–188
Markus ML (2001) Towards a theory of knowledge reuse: types of knowledge reuse situations and factors in reuse success. J Manage Inf Syst 18(1):57–93
McIvor R (2009) How the transaction cost and resource-based theories of the firm inform outsourcing evaluation. J Oper Manage 27(1):45–63
McKeen J, Smith H, Joglekar N, Balasubramanian PR (2002) Developments in practice: IT sourcing: building, buy or market. Commun AIS 9(8):120–135
McKinlay A (2002) The limits of knowledge management. New Technol Work Employ 17(2):76–88
Mills E (2002) Virtual recruitment. Nat Biotechnol Contents 20(8):853
Mudambi SM, Tallman S (2010) Make, buy or ally? Theoretical perspectives on knowledge process outsourcing through alliances. J Manage Stud 47(8):1433–1456
Nonaka I (1991) The knowledge-creating company. Harvard Bus Rev 69:96–105
Parry G, James-Moore M, Graves A (2006) Outsourcing engineering commodity procurement. Supply Chain Manage Int J 11(5):436–443
Quinn JB (1999) Strategic outsourcing: leveraging knowledge capabilities. Sloan Manage Rev 40 (4):9–21
Robinson HS, Anumba CJ, Carrillo PM, Al-Ghassani AM (2006) STEPS: a knowledge management maturity roadmap for corporate sustainability. Bus Process Manage J 12(6):793–808
Roos G (2005) Intellectual capital and strategy: a primer for today's manager. In: Coate P (ed) Handbook of business strategy, vol 6(1), Emerald Group Publishing Limited, UK, p 123–132
Rundquist J, Halila F (2010) Outsourcing of NPD activities: a best practice approach. Eur J Innov Manage 13(1):5–23
Sánchez AM, Pérez Pérez M, de Luis Carnicer P, Vela Jiménez MJ (2007) Teleworking and workplace flexibility: a study of impact on firm performance. Pers Rev 36(1):42–64
Schlosser F, Templer A, Ghanam D (2006) How human resource outsourcing affects organizational learning in the knowledge economy. J Labour Res 27(3):291–303
Schoenherr T (2012) The role of environmental management in sustainable business development: a multi-country investigation. Int J Production Economics 140(1):116–128
Stinchcombe AL (1965) Social structure and organizations. In: March JG (ed) Handbook of organizations. Rand McNally and Company, Chicago, pp 142–193
Storey J, Barnett E (2000) Knowledge management initiatives: learning from failure. J Knowl Manage 4(2):145–156
Tarn DDC, Chien-Chih E (2012) Knowledge outsourcing: a proposed model. Int J Manage Part 2 29(1):247–266
Van Kleef JAG, Roome NJ (2007) Developing capabilities and competence for sustainable business management as innovation. J Clean Prod 15(1):38–51
Wenger EC, McDermott R, Snyder WC (2002) Cultivating communities of practice: a guide to managing knowledge. Harvard Business School Press, Cambridge
World Commission on Economic Development (WCED) (1987) Our common future. Oxford University Press, Oxford

Wu F, Li HZ, Chu LK, Sculli D (2005) An outsourcing decision model for sustaining long-term performance. Int J Prod Res 43(12):2513–2535

Xia Y, Tang TL-P (2011) Sustainability in supply chain management: suggestions for the auto industry. Manage Decis 49(4):495–512

Zack MH (1999) Developing a knowledge strategy. Calif Manage Rev 41(3):125–145

The Role of Culture in a Sustainable Built Environment

Alex Opoku

Abstract The issue of the role of culture in a sustainable built environment is becoming more important due to the growing impact of the built environment in achieving society's sustainable development. Society faces environmental challenges, including climate change and urban population growth and responding to these interconnected challenges is at the centre of society's concerns for its future. This chapter aims to examine literature to advance this emerging area of research by mapping the linkages between the concept of culture, sustainable development and the built environment. A review of a wide-range of relevant literature relating to culture, sustainability and the built environment is presented. This review focused on relevant publications in academic literature that provide evidence of the role of culture in a sustainable built environment. The review of literature highlights the increasing recognition of culture as powerful and important aspect in fostering economic, social and environmental dimensions of development. Culture is a key element in the concept of sustainable development as it frames people's relationships and attitudes towards the built and the natural environment. Sustainable development is an integral part of the society and culture; affecting all aspects of operations in the built environment. This Chapter provides an in-depth insight into the contribution of culture and places culture as the possible fourth dimension of sustainability and an integral part of environmental, economic and social dimensions of sustainable development.

Keywords Culture · Cultural sustainability · Sustainable development · Sustainable built environment

A. Opoku (✉)
London South Bank University, London, UK
e-mail: opokua@lsbu.ac.uk

A. Chiarini (ed.), *Sustainable Operations Management*,
Measuring Operations Performance, DOI 10.1007/978-3-319-14002-5_3

1 Introduction

Cultural sustainability examines ways to improve human lives and how to leave a practical legacy for future generations. Society cannot continue with the current attitude towards resource exploitation if we have the aspiration to provide future generations with the ability to meet their own needs. Cultural value shape the society's way of life and therefore has the potential to bring about the change of attitudes needed to ensure the achievement of sustainable development. Culture contributes to the building of lively cities and communities where people can live, work and play a major role in supporting social and economic well-being. Hawkes (2001) argue that a sustainable society depends upon the vitality of a sustainable culture and cultural action arising from the communities as a result of their well-being and shared sense of meaning and purpose of its members. Cultural sustainability involves efforts to preserve the tangible and intangible cultural elements of society in ways that promote environmental, economic, and social sustainability (NZMCH 2006; Duxbury 2012). There is currently a paradigm shift in the worldview toward sustainability to include cultural sustainability in ways that do not damage our ecosystem, environment, and social well-being. Hawkes (2001) recognises that, the three dimensions of economic growth, social inclusion and environmental balance no longer reflect all the dimensions of our global societies.

The world is not only facing economic, social, or environmental challenges: creativity, knowledge, diversity, and beauty are the unavoidable basis for dialogue for peace and progress as these values are intrinsically connected to human development and freedoms (UCLG 2010). Culture and values are significant components of society, but each culture's values are not static; they evolve over time, thus first shifting society's view of what defines sustainability and then creating paradigm shifts in the worldview on sustainability (Fithian and Powell 2009). The Director-General of UNESCO (*Irina Bokova*) declared in 2012 that culture is what makes us who we are, providing answers to many of the challenges we face today and that we must do far more to place culture at the heart of the global sustainability agenda (UNESCO 2012a; Hayashi et al. 2013). The introduction and the integration of culture into the social, environmental, and economic dimensions of sustainability are vital in delivering a more holistic approach to sustainable development (Scammon 2012). Culture supports and connects with the other three pillars of sustainability; social, economic, environmental. For instance economic sustainability includes the evaluation of maintenance and operational costs of buildings in specific cultures while environmental sustainability supports the continuous use of historic buildings; avoiding demolition of historic buildings means waste materials to landfills are eliminated.

The cultural dimension of sustainability creates solid bridges with the other three dimensions of development and is compatible with each of them (UCLG 2010). The application of cultural sustainability can minimize the use of natural capital through resource management; improve tangible social capital such as public facilities and infrastructure; and strengthen economic capital by getting more out of renewable resources. The global built environment requires urgent improvement in energy and environmental performance (Grierson 2009) and this could only be achieved through a cultural change. The rest of the chapter is structured as follows; literature review on culture and sustainable development is described next. Then the literature on the contribution of culture on sustainable built environment is presented. Finally, literature findings and conclusions on the role of culture on a sustainable built environment are discussed.

2 Research Approach

A research approach is underpinned and bounded by the understanding and perspective of the researcher. Creswell (2007) stressed the importance of illustrating the research approach as an effective strategy to increase the validity of social research. Mackenzie and Knipe (2006) argue that methodology is the overall approach to the proposed research linked to the paradigm or theoretical framework to be used while the method refers to systematic modes, procedures or tools used for collection and analysis of data. The literature review provides a comprehensive critical analysis of the current thinking, from both an industrial and an academic perspective in the field of culture and sustainability in the built environment. The literature was considered in the context of theory so that objective evaluation could take place (Fellows and Liu 2003). The literature review examined literature on culture, sustainable development and the link between culture and sustainability in the built environment. This literature review aims to clearly identify the knowledge gap on the role of culture in a sustainable built environment. A critical review of literature developed an understanding of the structure of the subject, identifying what is known from what is unknown, establishing the significance of the problem under study, seeking new lines of enquiry and ascertaining the key variables essential to the issue (Randolph 2009). The chapter therefore presents a review of literature on the role of culture in a sustainable built environment.

3 Culture and Sustainable Development

Cultural sustainability is equally as important as economic, social and environmental dimensions of sustainable development and should be included as one of the dimensions supporting sustainability in a holistic approach. Sustainable development and culture are connected as culture describes society's understanding and appreciation of the natural resources and therefore plays an essential role in the promotion of economic progress in a fair society. A desired sustainable society's focuses on the three dimension of social equity and justice, environmental responsibility, and economic viability are essential but inadequate to support the future survival and well-being of humankind and therefore a fourth dimension (cultural development) is required (Hawkes 2001). Cultural heritage acts as a possible source of employment, social capital, energy savings and cultural diversity in attempt to pursue social, environmental and economic issues of sustainability. Continuous cultural heritage can be achieved when it is used as a tool for economic development because the acquired development will in turn ensure the sustainable development (Gunay 2008).

3.1 Sustainable Development

Sustainable development has been defined in many ways but the most commonly accepted definition of sustainable development is:

> Meeting the needs of the present without compromising the ability of future generations to meet their own needs… A process of change in which the exploitation of resources, the direction of investments, the orientation of technological development, and institutional change are all in harmony and enhance both current and future potential to meet human needs and aspirations (Brundtland 1987: 43).

However, Opoku and Ahmed (2013: 141) provide an alternative definition that addresses the concept of need and human behaviour as:

> The adjustment of human behaviour to address the needs of the present, without compromising the ability of future generations to meet their own needs.

The definition of Brundtland (1987) aims to be more comprehensive and addresses the key concept of needs while Opoku and Ahmed (2013) put emphasis on human behaviour in an attempt to meets their needs. The transition towards achieving sustainable development require changes in human behaviour, values and attitudes that will meet human needs.

Sustainable development balances environmental resource protection, social progress and economic growth and stability now and for the future. A vision of sustainable development with three dimensions namely: economic growth, social inclusion and environmental balance was consolidated at the Rio de Janeiro Earth Summit of 1992 as the paradigm of sustainable development (UCLG 2010). However, it is believed that these dimensions alone cannot possibly reflect the complexity of current society. There is now a call from agencies/institutions such as UNESCO, the World Summit on Sustainable Development, and researchers to include culture in the sustainable development model, since culture ultimately shapes what we mean by development and determines how people act in the world (Fithian and Powell 2009; UCLG 2010). Over the past decade, voices from different parts of society, from international institutions to academia, began to question the validity of the current definition of sustainable development. They state that economic growth, social inclusion and environmental balance, no longer reflect all the dimensions of our global societies and propose the addition of the dimension of culture to policies aiming at sustainable development (UCLG 2010). A Culture of peace fosters development and social cohesion and sustainable development could easily be achieved when there is the involvement of the local population at the centre of the development efforts. Culture is emerging as the fourth pillar of the sustainability model (Sustainable Development = Environment + Economy + Social + Culture). Integrating culture within development efforts is crucial to tackle a large number of these global challenges more effectively and sustainably. This model recognizes that a community's vitality and quality of life is closely related to the vitality and quality of its cultural engagement, expression, dialogue, and celebration (Fithian and Powell 2009). There is more to sustainability than environmental practices, economic growth, and equitable social services. Sustainability also includes a community's values and cultural heritage.

3.2 Cultural Sustainability

Culture play a role in defining human needs and interest making it critical to sustainable development. Cortese (2003) argues that a cultural shift is required to achieve a more sustainable society. Culture is central to society and defines people's attitudes and goals. It informs our understanding of development and controls our behaviour in the communities in particular and society in general. Agenda 21 for Culture (2004) places culture at the heart of the sustainable development processes. Parodi (2011) describes culture as an aim and a condition of sustainable development while Hauser and Banse (2011) believe that the relation between culture and sustainability is very rich. Whiles Opoku and Fortune (2011) believe that organizational leadership plays a vital role towards achieving sustainability, Taylor (2009) argues that efforts to change individual organizational leadership behaviour and attitude towards a more sustainable organization, without a parallel effort to change the culture within these organizations in which they operate will achieve little or no results. We need to develop a society where sustainable behaviour is a culture of the people.

There has been a growing recognition that culture and the environment function as parts of a single system and that they cannot be understood separately. Culture can be seen as acting driver of sustainable development since culture-related initiatives have direct impact in achieving sustainable development (NZMCH 2006; Fithian and Powell 2009). Culturally embedded processes in diverse settings play a significant role in sustainable Development efforts. Whenever cultural issues are considered in sustainability, it has been niched under the umbrella of social sustainability, however Hawkes (2001) advocates for a sustainability model that incorporates four interlinked dimensions of environmental responsibility, economic health, social equity, and cultural vitality (well-being, creativity, diversity and innovation) as illustrated in Fig. 1.

The four-dimension model of sustainability provides a cultural lens in evaluating the impacts of the environmental, economic, and social strategies being implemented (Fithian and Powell 2009). Nurse (2006) goes further by arguing that

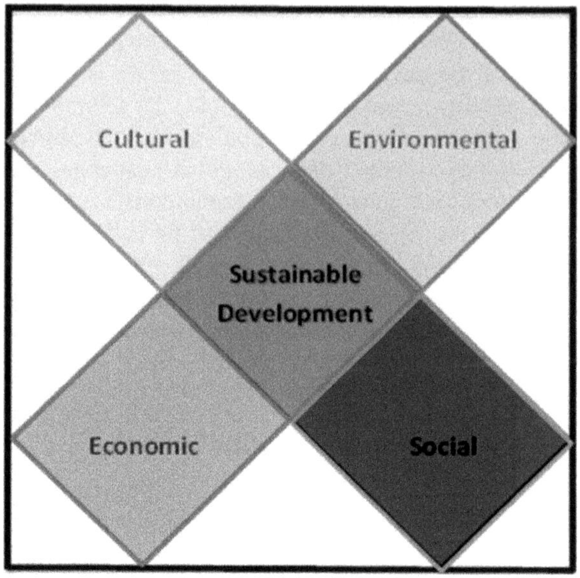

Fig. 1 Proposed four-dimension model of sustainability (*Source* Author 2014)

culture should be the central dimension of sustainable development and should be integrated fully into the economic, social, and environmental dimensions. He believes that people's identities and signifying systems shape how the environment is viewed. Cultural sustainability supports the other three dimensions of sustainability; social, economic, environmental through the re-use and recycling of resources on which energy has already been spent (NZMCH 2006).

Culture should be viewed not just as an additional dimension of sustainable development along with environmental, economic and social objectives because peoples' identities, signifying systems, cosmologies and epistemic frameworks shape how the environment is viewed and lived in (Fithian and Powell 2009).

The cultural dimension of sustainability is becoming increasingly present in the deliberations on sustainable development, but there are still some confusion regarding the definition of the term culture (Dallaire and Colbert 2012). Culture shapes what we mean by development and determines how people act in the world. Culture shapes what we mean by development and determines how people act in the world (Nurse 2006). Culture has been described as a set of unique spiritual, material, intellectual and emotional features of a society or a social group (Hayashi et al. 2013). Culture is both an instrument for decision-making and implementation as well as the end result of those policies and of the decisions implemented. Culture is about creativity, heritage, knowledge and diversity. These values are intrinsically connected to human development and freedoms (UCLG 2013). Integrating cultural aspects into research, decision-making and policy frameworks for environmental sustainability is an important challenge when addressing the cultural and social consequences of ecosystem change. Culture as a sector embraces tangible and intangible heritage, cultural and creative industries and cultural infrastructures, has made towards achieving sustainable development, as evidenced in terms of poverty alleviation, social inclusion and environmental sustainability (UNESCO 2012a).

Culture is a dynamic reality and constantly shifting process that plays an important role in sustainable development (Hawkes 2001). Local and indigenous knowledge systems and environmental management practices provide valuable insight and tools for tackling ecological challenges, preventing biodiversity loss, reducing land degradation, and mitigating the effects of climate change (UNESCO 2012a). Culture fosters economic growth, helps individuals and communities to expand their life choices, improving the resilience of social-ecological systems and adaptation to change. Duxbury and Jeannotte (2010) describe culture as capital (both tangible and intangible), a way of life (a system of social relations), a vehicle for sustainable values and creative expression providing insights on sustainability. The integration of culture into sustainable development strategies and policies advances a human-centered and inclusive approach to development in addition to serving as a powerful socio-economic resource. Culture is transversal and cross-cutting concern and, as such, affects all the dimensions of development (UNESCO 2012a). Culture is linked to the economy through income generation and employment, to social programs that deal with poverty, equal rights, and civic engagement; and culture is linked to the environment through the use of cultural capital to raise environmental awareness and responsibility (UCLG 2013). The contribution of culture to a sustainable built environment is discussed in the next section.

4 Culture and Sustainable Built Environment

Achieving a more sustainable society requires the redefinition of the relationship between the built and natural environments. The built environment consumes lots of energy and resources whiles generating large amount of waste. The solution to this challenge is the design and construction of improved built asset supported by appropriate management tools and regulatory frameworks that addresses sustainable development issues (Grierson 2009). A typical example of such tool or regulatory framework is the adoption of BRE Environmental Assessment Method (BREEAM) which sets the standard for best practice in sustainable building design and measure building's environmental performance. BREEAM is most widely used method of assessing, rating and certifying the sustainability of buildings and provides clients, developers, designers etc. a prove of the environmental credentials of their built assets.

4.1 The Built Environment

The built environment is literally a hub of economic activities, individuals and families, as well as society's cultural heritage. The built environment host economic activities, protects life and health, psychological and social welfare of its inhabitants, and sustain aesthetic and cultural values (Holm 2003). According to Bartuska (2007: 5), the Built Environment is defined as:

> Everything humanly made, arranged or maintained to fulfil human purpose (needs, wants and values) to mediate the overall environment with results that affect the environmental context.

However, Moffatt and Kohler (2008: 249) describe the Built Environment as:

> The manmade surroundings that provide the setting for human activity, ranging from the large-scale civic surroundings to the personal places.

The built environment involves human and cultural issues (why people build), environmental issues (natural and built context) and technological issues (materials, energy and financial resources, methods, and systems required to establish interrelationships and construct the built environment (Bartuska 2007). Building for sustainable development involves using design and construction methods and practices, which strive for integral quality (including economic, social and environmental performance) in a very broad (or holistic) way (Grierson 2009). The built environment is not just about buildings, infrastructure and transport but it also includes the interaction of people in the local communities and their cultural experiences (New Zealand Ministry for Environment 2009). An efficient sustainable built environment that includes economic and socio-cultural sustainability supports local economy (Allen 2009).

Hallsal (2011) explains that, the built environment is the physical human created surroundings such as roads, bridges, and building structures, contrasting that of the natural environment. The connections between components of the built environment, such as individual buildings, transport systems, urban landscapes and other infrastructure are important. The built environment revolves around communities full of cultural experiences that connect with people and the interaction of these factors creates a stronger and a more sustainable community with increased psychological well-being Jenkin and Pedersen (2009) believe that the built environment does not only consist of buildings, infrastructure and transport but it also includes human community, cultural experiences and interactions of people. The scope of the built environment consists of the following interrelated components: products, interiors, structures, landscapes, cities, regions, and Earth (Bartuska 2007). A sustainable built environment should aim at reducing environmental impacts in terms of energy, carbon, waste or water; this will involve creating a built environment that produces more than it consumes with environmental, social, cultural and economic benefits (Jenkin and Pedersen 2009). Cultural aspect of sustainability should be considered when formulating policies, developing and implementing development plans and programs for the built environment

(Jeannotte and Andrew 2012). Local and community needs (cultural environment) should be assessed and properly linked to the built asset project objectives to achieve a sustainable society (CIDA 1997).

4.2 The Role of Culture in a Sustainable Built Environment

Cultural sustainability and the connection with the built environment involves the development of building forms, materials and construction methods resonant with the local culture and expertise (Williamson et al. 2003). Organizations can contribute to achieving sustainability by reducing their negative impact on the natural environment while increasing the whole life of built asset in the built environment (Poston et al. 2010). Culture plays an important role as a non-renewable resource that is a vital part of our cities; the adaptive re-using of the historic urban environment adds to the quality of life of their inhabitants by strengthening their sense of belonging, social cohesion and providing a pleasant environment that mitigates excessive urbanization (UNESCO 2012b). The problem of climate change and the impact on a sustainable built environment is no more a question of science but rather a global issue and dealing with these issues involves different stakeholders of different socio-cultural perspectives (Holm 2003). This means that tackling the problem of climate change and the effects on the built environment require the integration of sustainable development education that can improves society's behaviours and attitudes to resources. Culture enables environmental sustainability at various levels; through the intrinsic links between cultural diversity and biodiversity, its influence on consumption patterns, and its contribution to sustainable environmental management practices as a result of local and traditional knowledge. Grierson (2009) explain that respect for all people, demonstrated through the provision of a healthy, functional, accessible and attractive built environment, is vital in promoting social sustainability. Culture can be an influential driver for sustainable development with broader social, economic and environmental impacts on the society as cultural heritage and cultural infrastructure can serve as strategic tools for revenue generation. Many cities across the world today use cultural heritage to encourage sustainable urban development and communities.

Culture contributes to environmental sustainability by conserving the historic environment and promoting creative design. Cultural heritage and creativity play a supporting role in achieving urban sustainability and facilitate a better fit between goals and processes and local conditions in development projects from health to education and infrastructure (Hosagrahar 2013). For example, Beynon (2010) observed that cultural sustainability can support the other dimensions of sustainability when buildings get used for new purposes, perhaps offering a socially sustainable benefit to low-income residents. Using a building for new purposes may offer an economic benefit because funds are not needed to build new; and it may also offer environmental benefits because building materials are not disposed as waste to landfills. Lewis (2012) argues that it takes 35–50 years for an energy-efficient new home to recover the carbon expended in its construction. Historic houses are built from brick, plaster, concrete and timber, or even from mud-brick, which are among the least energy-consuming materials.

The built environment is an integral part of spatial construction, and the history of building is a history of identity and therefore any alterations to buildings, involve elements and forms that relate to particular cultural and societal patterns (Beynon 2010). The cultural heritage is a powerful asset for inclusive economic development, since it attracts investment and ensures green, locally-based, stable and decent jobs relating to a wide range of sustainable activities in areas such as conservation, construction, (Hosagrahar 2013). It is therefore important that culture is placed at the centre of our development strategies because of the role culture plays in framing human behaviour and their relationship to others (UNESCO 2012b). Culture contributes to sustainable built environment through the protection of historic buildings, sustainable urban retrofit schemes, regeneration/urban renewal programmes, sustainable architecture and smart/sustainable cities within the realms of the community's identity. A model of the role of culture in a sustainable built environment developed from the literature review is presented in Fig. 2.

Culture therefore has an essential role to play in the development of a sustainable built environment that includes infrastructure that supports neighbourhoods and cities. Sustainable building design should meet the end users social and cultural needs while preserving the environment for future generation (Al-Jamea 2014).

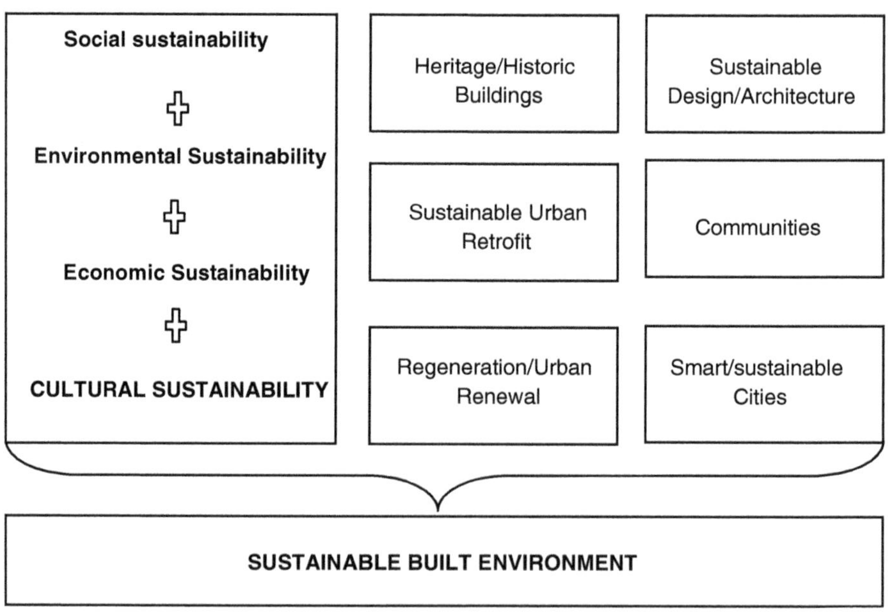

Fig. 2 Culture and sustainable built environment framework (*Source* Author 2014)

4.3 Cities, Culture and Sustainable Development

Culture has been the underdeveloped component of both conceptual and planning frameworks for long-term community well-being and sustainability. Historic towns and the historic parts of cities are valuable for their uniqueness and identity (Duxbury 2012). Creativity in urban areas has increasingly become part of the culture-led redevelopment of urban areas in order to bring about improved economic returns and increased competitiveness between cities (Hosagrahar 2013). Landscapes and the components of them have always strongly influenced local cultural practices, values and beliefs. In urban areas, initiatives such as sustainable architecture, ecological design, and eco-arts can stimulate creativity and innovation by linking ecological concerns, well-being and aesthetics (Duxbury et al. 2012). Cultural considerations should be integrated into sustainable urban and community planning through strong cultural programs that link intercultural realities and the local environment. Andrew et al. (2011) argue that, more research is needed to increase the understanding of the concept of integrating culture into sustainability planning and policy development.

Today, many cities use cultural heritage and cultural events and institutions to improve their image, stimulate urban development, and attract visitors as well as investments (UNESCO 2012a). The cultural heritage is linked to the lives of communities and is fully integrated into social, economic and environmental processes, making it an integral part of people's daily experience (Hosagrahar 2013). Maintaining the cultural heritage is also important in addressing risks relating to natural and human-made disasters. Whiles recent city planning paradigms are focusing on achieving sustainable cities through sustainability planning initiatives such as the adoption of sustainability goals for individual communities and regions, cultural considerations has not been integrated generally into planning; even though culture has been recognized in urban and community planning contexts (Duxbury and Jeannotte 2010; Duxbury 2012). When urban development policies and strategies are driven by heritage (culture), it helps contributes to alleviating and adapting to climate change. For example, in UK, the embodied energy in the construction of a building is estimated at 15–30 times the annual energy use; which means that it is more sustainable to conserve the existing fabric of our built environment constructed with traditional methods, local materials and skills than demolishing and reconstructing (Lewis 2012; UNESCO 2012b). Current urbanization policies often ignore the importance of cultural heritage preservation and promotion and the great potential of creativity in addressing social, environmental and economic urbanization challenges (Hosagrahar 2013; Duxbury et al. 2012).

Cultural heritage, particularly cultural landscapes and historic cities, can make a significant contribution to environmental sustainability. Lewis (2012) argues that replacing a historic building with a new one involves the higher consumption of energy. Demolishing a historic building means wasting the energy incorporated into that building, and it requires further energy to take the materials to a landfill and even more to reconstruct a new building. Cultural knowledge in dealing with the risks of floods, mudslides, droughts, as well as earthquakes, is valuable in helping to make communities resilient. In addition, local building technologies and the use of local materials in heritage structures and the knowledge of ecosystem management that is inherent in their responsiveness to climate, topography, and available natural resources are all significant as sources of knowledge and offer a diversity of solutions for particular conditions (Duxbury 2012; Hosagrahar 2013).

The conservation of historic cities is an effective strategy in reducing CO_2 emissions when compared to their replacement by new buildings. New construction projects have more damaging impacts in short to mid-term, however conserving an existing building rather than demolishing and reconstructing reduces the demand for new and saves original energy and CO_2 investment (Lewis 2012). Historical preservation and reuse of old buildings is the most sustainable practice in the built environment today and a crucial strategy towards encouraging cultural sustainability and counter the effects of global warming.

5 Discussions

The protection, promotion and maintenance of cultural sustainability is vital for sustainable development that benefit present as well as future generations. Culture is integral part of sustainable development; however the key challenge is the understanding of the relationships between culture, sustainability and development. Culture indeed has role to play towards achieving a sustainable built environment because culture is a key element in the concept of sustainable development as it frames people's relationships and attitudes towards both the built and the natural environment. Cultural value shape peoples way of life and therefore has the potential to bring about the change in behaviour and attitude needed to ensure the achievement of sustainable development. It is argued that economic growth, social inclusion and environmental balance, no longer reflect all the dimensions of our current world and a sustainable society could not be achieved without a cultural dimension. Sustainable development and culture are connected; cultural sustainability is equally as important as the economic, social and environmental pillars of sustainable development and should be included as one of the pillars of sustainability. Despite the above essential role of culture in a sustainable society, the cultural dimension of sustainability has not been fully accepted in the academic. Especially in construction management research, little or nothing has been written on the link between culture, sustainability and the built environment. Currently there is a growing recognition that culture and the environment function as parts of a single system and that they cannot be understood separately. Cultural sustainability also supports the other three dimensions of sustainability; social, economic, environmental and is becoming increasingly present in the discussions on sustainable development.

A sustainable built environment adopts design, construction methods and practices that make use of local expertise and experience. Our built environment operates in a society and local communities full of cultural experiences that connect with people. Even though current urban policies often ignore the importance of preserving and promoting cultural heritage, it is important that cultural considerations are integrated into sustainable urban and community planning. Culture contributes to the building of lively cities and communities where people can live and work playing a major role in supporting social and economic well-being. Culture contributes to sustainable built environment through the protection of historic buildings, sustainable urban retrofit schemes, regeneration/urban renewal programmes, sustainable architecture and smart/sustainable cities within the realms of the community's identity. Cultural aspect of sustainability should therefore be considered when formulating policies towards sustainable the built environment. The development of policies towards sustainable development at all levels of organization and sustainability should integrate cultural beliefs and experiences. Culture should be placed at the centre of our development strategies because culture plays an important role in framing human behaviour and their relationship to others.

6 Conclusions

The cultural dimension of sustainability should be integrated into development policies and the idea of culture as the fourth dimension of sustainability should be promoted internationally. The chapter recognizes the increasing need to incorporate culture into sustainable development plans and strategies. Cultural sustainability projects benefit the environment through the preservation of cultural capital such as buildings that retain a community's heritage. The chapter argues that culture should be viewed not just as the fourth dimension but as the central dimension of sustainable development.

It is noted from the literature review that, the absence of culture from the core goals and methods used to effect sustainable development was due to the difficulties in measuring its concrete contribution to development; however this challenge has been resolved in recent years through the multidimensional measures of human well-being. This chapter therefore recommends that particular attention should be paid to measuring the quantitative and qualitative contributions of culture to sustainable development by establishing clear indicators. Cultural well-being occurs when society and individuals are provided social activities that promote cultural capital. Cultural sustainability involves efforts to preserve the tangible and intangible cultural elements of society in ways that promote economic sustainability. The development and the promotion of a sustainable built environment should be carried out in an environmentally, economically, and socially beneficial manner with culture as the overarching dimension of sustainable development. Cultural sustainability supports and connects the other three dimensions of social, economic and environmental sustainability. The integration of the cultural dimension of sustainability should therefore be promoted through education and public policies towards a sustainable built environment. The implementation of policies, development plans, and strategies for sustainable cities and communities around the globe should consider cultural issues because cultural content shapes and communicates the identity, values and hopes of a society.

In conclusion, sustainability efforts that once were characterized by environmental, social, and economic issues only now see the inclusion of culture as a holistic benefit to sustainable development. Finally, the literature review provides the opportunity from an academic perspective to empirically further investigate the contribution of culture to the understanding of sustainable development and the built environment in future research.

References

Agenda 21 for Culture (2004) An undertaking by cities and local governments for cultural development. Available at: www.barcelona2004.org/…/t_portoalegreeng.pdf. Accessed 12 Mar 2014

Allen A (2009) Sustainable cities or sustainable urbanisation? 'Palette'-UCL's J Sustain Cities. Available at: www.ucl.ac.uk/sustainable-cities. Accessed 10 Sept 2014

Al-Jamea M (2014) Towards social and cultural sustainability in the designs of contemporary Saudi houses. Int J Sustain Hum Dev 2(1):35–43

Andrew C, Duxbury N, Jeannotte MS et al (2011) Culture in sustainable communities – a Canada-Europe comparative research project. Paper presented at Policy Research Speakers Series, Department of Canadian Heritage, Gatineau, Canada, 13 October

Bartuska TJ (2007) The built environment: definition and scope. In: McClure WR, Bartuska TJ (eds) The built environment: a collaborative inquiry into design and planning, 2nd edn. Wiley, Oxford, pp 3–14

Beynon D (2010) Defining cultural sustainability in multicultural built environments. Int J Environ, Cult, Econ Soc Sustain 6(5):255–266

Brundtland GH (1987) Our common future: report of the world commission in environment and development. Oxford University Press, Oxford

CIDA (1997) Our commitment to sustainable development: the strategy of the Canadian International Development Agency. Canadian International Development Agency, Ottawa

Cortese AD (2003) The critical role of higher education in creating a sustainable future. Plan High Educ 31(3):15–22

Creswell JW (2007) Qualitative inquiry and research design: choosing among five approaches, 2nd edn. Sage, Thousand Oaks

Dallaire G, Colbert F (2012) Sustainable development and cultural policy: do they make a happy marriage? ENCATC J Cult Manag Policy 2(1):6–11

Duxbury N (2012) Cities, culture and sustainable development, Summer Lab'12: art, climate change and community development. Torres Vedras, Portugal, 5 Sept 2012. http://www.ces.uc.pt/myces/UserFiles/encontros/1097_DUXBURY_Torres_Vedras_05.09.2012.pdf. Accessed 24 Nov 2013

Duxbury N, Jeannotte MS (2010) Culture, sustainability and communities: exploring the myths, Oficina do CES Nr. 353, Universidade de Coimbra. A paper presented at the 6th international conference on cultural policy research, Jyväskylä, Finland, 24–27 Aug 2010

Duxbury N, Cullen C, Pascual J et al (2012) Cities, culture and sustainable development. In: Anheier HK, Isar YR, Hoelscher M et al (eds) Cultural policy and governance in a new metropolitan age. The cultures and globalization series, vol 5. Sage Publications Ltd, London, pp 73–86

Fellows R, Liu A (2003) Research methods for construction students, 2nd edn. Blackwell publishing, Oxford

Fithian C, Powell A (2009) Cultural aspects of sustainable development. Student paper for sustainable design seminar/studio, fall 2009, seminar in sustainable architecture, University of Texas School of Architecture. Available via: http://repositories.lib.utexas.edu/bitstream/handle/2152/11698/5-Fithian_Powell-Cultural_Aspects_of_Sustainable_Development.pdf?sequence=2. Accessed 20 Dec 2013

Grierson D (2009) Towards a sustainable built environment. CIC Start Online Innovation Rev 1:70–77

Gunay Z (2008) Neoliberal urbanism and cultural heritage. 44th ISOCARP congress, 2008

Hallsal P (2011) Culture, Sustainability and the built environment: revisiting heritage and vernacular in economic decline. Int J Glob Built Environ Rev (GBER) 8(1):71–73

Hauser R, Banse G (2011) Culture and culturality: approaching a multi-faceted concept. In: Parodi O, Ayestaran I, Banse G (eds) Sustainable development-relationships to culture, knowledge and ethics. KIT Scientific Publishing, Karlsruhe

Hawkes J (2001) The fourth pillar of sustainability: culture's essential role in public planning. Common Ground Press, Melbourne

Hayashi N, Boccardi G, Al Hassan N et al (2013) Culture in the post-2015 sustainable development agenda: why culture is key to sustainable development. http://www.unesco.org/new/fileadmin/MULTIMEDIA/HQ/CLT/images/Post2015SustainableDevelopmentAgendaENG.pdf. Accessed 20 Nov 2013

Holm FH (2003) Towards a sustainable built environment prepared for climate change? A paper presented at the global policy summit on the role of performance-based building regulations in addressing societal expectations, international policy, and local needs. National Academy of Sciences, Washington, DC, 3–5 Nov 2003

Hosagrahar J (2013) Built Heritage and sustainable development: a case study from South Asia. In: Proceedings of The Hangzhou international congress, session 3a-culture's contribution to achieving sustainable cities, Hangzhou-China, 15–17 May 2013

Jeannotte MS, Andrew C (2012) Integrating culture in sustainable communities: canadian policy and planning approaches symposium on placing culture in sustainable development, Coimbra, Portugal, 15 May 2012

Jenkin S, Pedersen M (2009) Rethinking our built environments: towards a sustainable future. A research document. Ministry for the Environment, Wellington

Lewis JO (2012) Renovate or demolish/rebuild: what are the drivers? EuroACE Brussels, Renovate Europe Day, 11 Oct 2012. http://www.renovateeurope.eu/uploads/REDay2012%20ppts/REDay2012_Workshop_Owen_Lewis.pdf. Accessed 12 Dec 2013

Mackenzie N, Knipe S (2006) Research dilemmas: paradigms, methods and methodology. Issues Educ Res 16(2):193–205

Moffatt S, Kohler N (2008) Conceptualizing the built environment as a social-ecological system. Build Res Inf 36(3):248–268

New Zealand Ministry for the Environment (2009) Rethinking our built environment: towards a sustainable future. New Zealand Government

Nurse K (2006) Culture as the fourth pillar of sustainable development. Prepared for Commonwealth Secretariat, UK

NZMCH (2006) Cultural well-being and local government report 1: definition and context of cultural well-being. New Zealand Ministry for Culture and Heritage (NZMCH), Wellington

Parodi O (2011) Three steps towards a culture of sustainability. In: Parodi O, Ayestaran I, Banse G (eds) Sustainable development-relationships to culture, knowledge and ethics. KIT Scientific Publishing, Karlsruhe, pp 76–92

Poston A, Emmanuel R, Thomson C (2010) Developing holistic frameworks for the next generation of sustainability assessment methods for the built environment. In: Egbu C (ed) Proceedings 26th annual ARCOM conference. Association of Researchers in Construction Management, 6–8 Sept 2010, Leeds, UK, pp 1487–1496

Opoku A, Ahmed V (2013) Understanding sustainability: a view from intra-organizational leadership within UK construction organizations. Int J Archit Eng Constr 2(2):133–143
Opoku A, Fortune C (2011) The implementation of sustainable practices through leadership in construction organizations In: Egbu C, Lou ECW (eds) Proceedings 27th annual ARCOM conference, 5–7 Sept 2011, Bristol, UK, Association of Researchers in Construction Management, pp 1145–1154
Randolph J (2009) A guide to writing the dissertation literature review: practical assessment, research and evaluation. Pract Assess Eval 13(14):1–13
Scammon D (2012) Sustainability and culture: how do they work together? LCC 480 senior seminars, 30 April 2012. Available at: http://www.academia.edu/1817961/Sustainability_and_Culture_How_do_they_work_together. Accessed 25 Sept 2014
Taylor C (2009) Building cultures for sustainability. Oxford Leadership J 1(1):1–3
UCLG (2010) Culture: fourth pillar of sustainable. UCLG policy statement, UCLG Committee on culture and world secretariat policy statement for Mexico City, United Cities and Local Governments (UCLG), Barcelona, 16 Sept 2010
UCLG (2013) The role of culture in sustainable development to be explicitly recognised. UN economic and social council (ECOSOC 2013), UCLG Committee on culture world secretariat of United Cities and Local Governments (UCLG), Barcelona
UNESCO (2012a) Culture: a driver and an enabler of sustainable development. UN system task team on the post-2015 development agenda, The United Nations Scientific, Educational and Cultural Organization (UNESCO). http://www.unesco.org/new/fileadmin/MULTIMEDIA/HQ/post2015/pdf/Think_Piece_Culture.pdf. Accessed 15 Dec 2013
UNESCO (2012b) The future we want: the role of culture in sustainable development. The United Nations Scientific, Educational and Cultural Organization (UNESCO). http://www.unesco.org/new/en/rio-20/culture-for-sustainable-development/. Accessed 15 Dec 2013
Williamson T, Radford A, Bennets H (2003) Understanding sustainable architecture. Spon Press, London

The Management of Environmental Performance in the Supply Chain: An Overview

Enrico Bracci and Mouhcine Tallaki

Abstract The Environmental Supply Chain Management (ESCM) has become a buzz-word nowadays. This is in part due to the increasing awareness of companies that the environmental performances need to be managed beyond the organizational boundaries. The present paper aims at contributing to the extant literature through a comparative analysis of some frameworks that attempted to connect the performance measurement with the ESCM. From a methodological point of view, the paper is based on a literature review using as keywords the environmental performance management system and supply chain. Three main frameworks for the performance measurement ESCM were analyzed, in details: the Beamon Model (Logistics Information Management 12(4):332–342, 1999), the Hervani et al. (2007) and the Balanced Scorecard variation proposed by Epstein and Wisner (Balanced Scorecard Report 3(3):8–11, 2001). In so doing, we suggest possible commonalities and differences, arguing the need for further research to develop and understand the way companies manage the environmental performance of their supply chain.

Keywords Supply chain · Environmental performance · Supply chain management

The paper can be attributed to the authors in the following manner: Sects. 1 and 2 to Mouhcine Tallaki, Sects. 3 and 4 to Enrico Bracci.

E. Bracci (✉) · M. Tallaki
Department of Economics and Management, University of Ferrara, Ferrara, Italy
e-mail: enrico.bracci@unife.it

M. Tallaki
e-mail: mouhcine.tallaki@unife.it

A. Chiarini (ed.), *Sustainable Operations Management*,
Measuring Operations Performance, DOI 10.1007/978-3-319-14002-5_4

1 Introduction

The Environmental Supply Chain Management (ESCM)[1] has become a buzz-word nowadays. This is in part due to the increasing awareness of companies that the environmental performances need to be managed beyond the organizational boundaries. The ESCM is derived from the integration of environmental management and supply chain management concepts (Srivastava 2007; Thun and Muller 2010; Testa and Iraldo 2010). The environmental management is based on the principles of prevention, environmental protection, transparency and participation. Its basic goals are the identification of environmental issues at the enterprise level, the adoption of best environmental technologies, the reduction of environmental impacts and the use of prevention systems to avoid and/or treat adequately the potential environmental hazards due to specific production activities processes.

The supply chain management, instead, is defined as *The systemic, strategic coordination of the traditional business functions and tactics across these business functions within a particular company and across businesses within the supply chain, for the purpose of improving the long-term performance of the individual companies and the supply chain as a whole* (Mentzer et al. 2001). This definition recognizes explicitly the strategic nature of coordination between trading partners and it explains the twofold objective of supply chain management that are improving the performance of a single organization and of the entire supply chain (Gunasekaran et al. 2004; Bratic 2011). The ESCM is a management approach that has attracted the attention of the scientific community only recently. Srivastava (2007) pointed out that this approach is gaining increasing attention and interest among researchers and practitioners engaged in supply chain management and operations. This is mainly due to the progressive deterioration of the environment, the decrease in raw material resources, increase of pollution levels and of to the current issues related to waste management. In addition to attention to the environment, the use of ESCM practices could lead to improve economic as well as that of the environment performance, this justifies their adoption by firms (Srivastava 2007).

However, the ESCM needs to be integrated with the other managerial systems and in particular the performance measurement system. This is relevant in order to support managers in the decision-making, in the evaluation of performance and to assess the efficacy and coherence of output with the strategy. In the literature ESCM and the performance measurement tend to be studied in a separate way, despite having several contact points and calls are made to study them in conjunction (Melnyk et al. 2014).

The present paper aims at contributing to the extant literature through a comparative analysis of some frameworks that attempted to connect the performance measurement with the ESCM. From a methodological point of view, the paper is

[1] In the literature ESCM is also referred as Green Supply Chain Management. We are going to use the two concepts (ESCM and GSCM) as synonymous.

based on a literature review using as keywords the environmental performance management system and supply chain. Three main frameworks for the performance measurement ESCM were analyzed, in details: the Beamon Model (1999), the Hervani et al. (2005) and the Balanced Scorecard variation proposed by Epstein and Wisner (2001). In so doing, we suggest possible commonalities and differences, arguing the need for further research to develop and understand the way companies manage the environmental performance of their supply chain.

The rest of the paper will be structured in the following fashion: the next section will present a literature review of the ESCM concept. The third section will analyze the literature related to the performance measurement in the supply chain before describing the three models proposed to measure the environmental performance, while the fourth section will compare and discuss the models. The paper will conclude with some indications for further research.

2 The Environmental Supply Chain Management

Given the relative novelty of the ESCM discipline, there is no exhaustive definition and nor a consensus on what are the set of practices that can be adopted by companies (Zhu and Sarkis 2004). The concept of ESCM is considered as part of the Sustainable Chain Management (SCM) discipline. Seuring and Muller (2008) defined the ESCM *as the management of material, information and capital flows as well as cooperation among companies along the supply chain while taking goals from all three dimensions of sustainable development (i.e. economic, environmental and social), into account which are derived from customer and stakeholder requirements, In sustainable supply chains environmental and social criteria need to be fulfilled by the members within the supply chain, while it is expected that competitiveness would be maintained through meeting customer needs and related economic criteria*. This definition frames the ESCM within the broader Sustainable Supply Chain Management. The latter has the objective to improve the social policies and the environmental and economic impact through the supply chain management, the exchange of information, knowledge and resources and the sharing of common objectives between the stakeholders involved.

Some authors proposed the concept of Green (or Environmental) Supply Chain Management (GSCM) (e.g. Srivastava 2007). The GSCM considers the environment in its traditional functions and processes. In particular those functions related to supply chain management, plan, source, make, deliver and return (based on the classification of the Supply Chain Council). Srivastava (2007) highlighted the importance of the new management approach and the existence of two others research topics on GSCM: the first focuses on green design, the second instead emphasizes on the green operations (see Fig. 1).

According to Thun and Muller (2010), the supply chain management implies a greater inter-organizational cooperation, the concept of the GSCM follows the same approach. Adding the prefix of "*green*" this concept includes, then, all the efforts

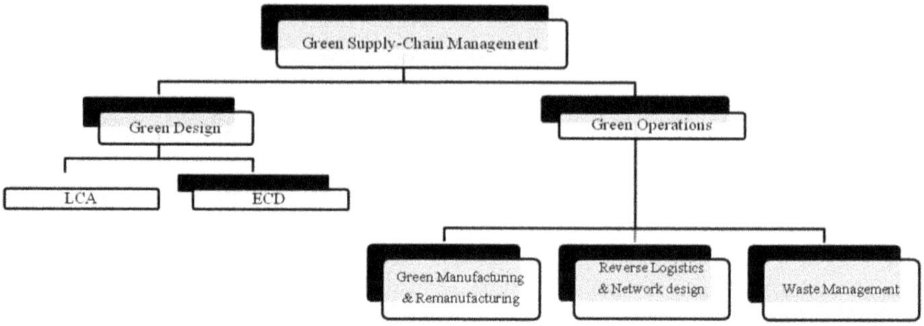

Fig. 1 Classification based on problem context in supply chain design. *Source* Authors' elaboration on data from Srivastava (2007)

that minimize the environmental impact of the company's products and processes. Thus, the key element of GSCM and ESCM alike is the integrated design of inter-organizational processes witch improve environmental efficiency. Hervani et al. (2005) defined the GSCM as the management of the effects of SCM on the environment with the goal to eliminate and reduce waste and environmental impacts (energy, emissions, chemicals substances and solid waste) along the supply chain. In particular, the GSCM is defined according to set of activities like green purchasing, green manufacturing and material management, green distribution, marketing and reverse logistics. Liu et al. (2011) consider the GSCM as an advanced system of Corporate Environmental Management. The latter aims to provide greener products and services through communication and cooperation. Handfield et al. (2005) proposed a definition of ESCM considering it as an environmental system that focused on relationships with suppliers and on the production of internal and external environmental information.

Therefore, there is not a single definition of ESCM, which sometimes is defined as GSCM. Some definitions show more than others the importance of appropriate relations with those involved along the supply chain, and other place attention on particular initiatives related to the environmental performance (Ninlawan et al. 2010). The initiatives and the extension of ESCM practices inside and outside the organization depend on many factors. They depend on the goals that are defined consistently with other business strategies, and also on sharing of these strategies between different organizations along the supply chain. As well as on the firm's ability to select, influence and control their business partners and then to design the supply chain. These initiatives, as has been identified, should be concentrated to make processes and products more environmentally friendly and to reduce the environmental impacts of the organization and the supply chain through strategies such as Green Design, Green Procurement (or Green Purchasing), Green Manufacturing (or Green Production), Green Logistics (or Green Distribution) and Reverse Logistics (Srivastava 2007).

3 ESCM and Performance Measurement Systems

The implementation of ESCM practices requires its formalization at the strategic level (Zhu et al. 2008). Because competitive strategies significantly influence the effectiveness of supply chain integration (Huo et al. 2014). In this way, it is possible to identify the objectives of reducing environmental impacts, in a perspective of supply chain, coherent with the other strategies of the company. This enables the new management philosophy to gain legitimacy in all levels of the organization. To do that it is necessary to have an unwavering support and commitment of the top management (Zhu et al. 2008).

The role of the supply chain should be considered by the companies in the formulation of its strategic plans. The use of outsourcing and de-verticalization strategies increase the importance of developing appropriate relationships with suppliers in order to achieve new and efficient forms of coordination. It is, thus, necessary to measure and define clear goals to be achieved, identify the means and the time to do so, identify and communicate the various responsibilities, modify the organizational structure and roles, implement systems that control the achievement of the objectives also through the choice of appropriate indicators of environmental performance and adopt a management control able to measure them. In addition, it will be interesting to balance various qualitative and quantitative, economic-financial and non-monetary measures, providing information at both management and operational level.

An ESCM incorporates the traditional mechanisms of management, including budgeting, performance measurement systems and risk management processes, which are commonly used in the management, but that are designed to treat and manage the specific problems and issues relating to the environment (Zhu et al. 2008). These tools are increasingly popular in the environmental strategies of the organizations to coordinate, monitor and manage the information, thus they represent an effective management control system (Arjaliès and Mundy 2013). If the environmental strategy is considered as a new dimension of the supply chain, the systems that manage the environmental strategy such as environmental management systems have to be properly extended and modified to support the new perspective embraced by the company. Some authors (i.e. Hervani et al. 2005) highlighted the need to design performance measurement systems integrated with the ESCM (Melnyk et al. 2014). However, the consideration of environmental issues at the strategic level, in particular within the supply chain perspective, involves a deep organizational change. The control systems needs to be designed and used to align the organizational behavior towards the achievement of new strategic objectives.

Management control systems is defined as "*the formal, information-based routines and procedures managers use to maintain or alter patterns in organizational activities*" (Simons 1995). Their use could be expressed through control levers such as diagnostic, interactive, belief and boundary levers. The use of these levers together could facilitate the implementation and achievement of the strategic objectives of the organization (Arjaliès and Mundy 2013).The management control systems can be used as a means to implement, monitor and discuss the business strategy relating to Corporate Social Responsibility (CSR). This field of study has been extensively studied by the scientific community. Not equal attention was given to the activities of ESCM. Arjaliès and Mundy (2013) used the framework of the control levers to analyze the way in which 40 companies listed in France implement and manage strategies of CSR. The authors showed that, for example, the diagnostic processes are used to manage the key variables of outcome. The purpose of the authors was to define and measure key performance indicators of CSR strategy regarding the internal and external objectives. This purpose is achieved through the use of CSR reports that manage the activities at the operational level in relation to the CSR performance activities which considered as critical. According to Arjaliès and Mundy (2013) management control systems have the potential to contribute to the achievement of greater sustainability through processes that enable the innovation, communication, reporting and identification of opportunities and threats. These levers of control could also be used to achieve the objectives defined by ESCM strategies.

Performance measurement systems have a high importance within companies because they allow to manage the organization through the analysis of the results achieved. They also enable to identify the business areas that are below expectations, and also contribute to their subsequent improvement through the implementation of corrective actions. Melnyk et al. (2014) argued recently that performance measurement systems of the future will need to take particular attention to the new business and organizational models based on extended supply-chain and networks. They are also important to influence behavior, so these systems have to be designed to measure, along with others, the results of the ESCM initiatives and to guide the actions of managers in order to achieve strategies. The indicators (mainly environmental) that constitute the core of the system could also be used as interactive tools when they promote dialogue and constant comparison (Rodrigue et al. 2013). Several studies made calls to study performance measurement throughout the supply chain (Melnyk et al. 2014; Bititci et al. 2012; Gunasekaran et al. 2004). So, Supply chain management is considered as key strategic factor for increasing organizational effectiveness and for better realization of organizational goals such as enhanced competitiveness, better customer care and increased profitability (Gunasekaran et al. 2001). Generally, the adoption of ESCM practices leads to improved environmental performance and economic performance, which, in turn, positively impact on operational performance (Green et al. 2012). Several studies have analyzed the relationship between performance and ESCM. Rao and Holt (2005) identified that greening the different phases of the supply chain leads to an integrated green supply chain, which ultimately leads to competitiveness and economic performance. Zhu and Sarkis (2004), using empirical results from 186 respondents on GSCM practice in Chinese manufacturing enterprises, examined the relationships between GSCM practice and environmental and economic performance. The authors argued that GSCM practices tended to have win-win relationships in terms of environmental and economic performance. Then, being proactive on environmental issues can bring companies a great number of benefits that improve their competitiveness and firm performance (Femenias et al. 2013).

According to Handfield et al. (2005), to manage adequately the objective of improving the environmental performance at the level of the supply chain we have to link the business strategy with supply strategy for classes of materials, components and products. The alignment between supply chain strategy and environmental uncertainty is positively associated with supply chain management performance (Sun et al. 2009). The idea is to define, at strategic level, the importance of pursuing an improvement of the environmental performance of the organization and the supply chain. The strategic formulation would lead to identify the objectives and general principles that inspire the action of the company and that will be pursued in the long term. However, these strategic objectives related to the environment should be declined in strategies at the functional level of business-unit and then at the operational level. This deployment system (Fig. 2) of the company strategy, regarding the environmental variable and the supply chain, has to focus particularly on the environmental performance of suppliers to reduce their environmental impacts along the supply chain. The idea highlighted by the authors is that the company inherits through purchasing decisions, the environment impacts (and risks) that have occurred along its supply chain. The company could control

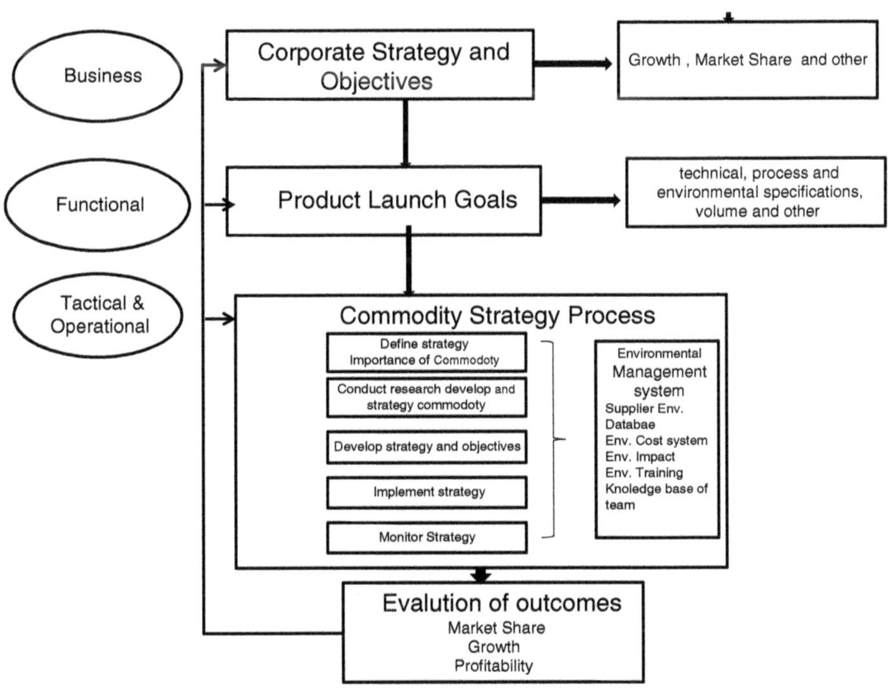

Fig. 2 Conceptual model of environmental role in supply chain decisions. *Source* Authors' elaboration on data from Handfield et al. (2005)

these impacts (with the aim to reduce them) through an appropriate process for selecting suppliers. Probably, the company that behaves in this way already adopts technologies and processes that are environmentally friendly. The interaction between companies and suppliers could potentially develop and implement more effective solutions to environmental challenges they are facing (Vachon and Klassen 2006). This occurs especially when there is a collaboration between companies and suppliers. Vachon and Klassen (2006) examined how the environmental collaborative activities impacts environmental performance. Environmental collaboration was defined by the authors specifically to focus on inter-organizational interactions between supply chain members, including such aspects as joint environmental goal setting, shared environmental planning, and working together to reduce pollution or other environmental impacts. The collaboration also implies cooperation to reduce the environmental impact associated with material flows in the supply chain (Bowen et al. 2001; Carter and Carter 1998).

The framework proposed by Handfield et al. (2005) is interesting because it integrates strategies of supply chain management with environmental management and it allows to highlight the links between ESCM and business strategy. The authors propose a system to achieve this objective by introducing the environmental strategy defined at the corporate at functional and operational levels of business unit. The objective is to ensure a consistency with all strategies at different levels. The operational level guarantees the achievement of a greater control of the environmental performance of their suppliers, because for each class of materials and products it has identified the critical level and then actions to better manage relationships with suppliers in order to reduce environmental impacts and improve environmental performance.

However, it is clear that companies, who engage in the implementation of ESCM practices in broad measure, and which also involve their business partners in the initiative to achieve greater environmental sustainability, need an integrated tools that allow them to guide the business activities through adequate measurement of different dimensions of business performance including environmental performance (possibly also suppliers to express the performance of the supply chain) and that provide an overview on what has been done towards achieving the objectives established (Franco-Santos et al. 2007). The PMS allows, if properly designed, to control the actions of the company by providing managers with the information that they need to conduct its business decision making (to lead) with reference to the environmental variable.

The criticisms of the traditional management control systems, which focus almost on measuring economic and financial performance, led to the development of advanced measurement and management control systems that consider simultaneously different dimensions of corporate performance (Kaplan and Norton 1992; Atkinson et al. 1997). Economic and financial performance are no longer sufficient to provide all relevant information to guide action of managers and in particular at operative level (although still important). Non financial measures include product quality or product characteristic (Perera et al. 1997) have non-monetary nature, they are sometimes associated only with the operational level, while they have also strategic importance (which is usually underestimated), they also impact profit and profitability. Their operational nature is critical because they are taken into account for decisions at the operational level, however, the new context requires that this part of indicators is integrated in the directional systems of information to provide multidimensional and balanced performance management systems. Companies have to shift attention from cost and efficiency-based performance measures to ones which capture and reflect strategies (Perera et al. 1997). The quality characteristic, flexibility and speed of decision making are increasingly important in the new competitive environment and therefore these dimensions must also be captured and reported to management even if they originate at the operational level. This allows to give a true picture of the current performance and future prospects of the business. Among these dimensions we can incorporate also environmental performance.

Given the new dimensions of the competitive environment, organizations should innovate the traditional performance measurement systems by identifying new performance measurement systems that meet the multidimensional information needs. Several authors agree on the importance of new systems of performance measurement, although some criticize how is vague the concept of PMS (Azevedo et al. 2013). In fact, the definition of PMS is very general and for this reason they have been developed different models proposed by various authors to deal with the problem of performance measurement in a concrete way, identifying indicators, measures and perspectives of analysis, that have to be appropriate and standardize for each company. The advanced performance management systems measure and report a series of multi-dimensional indicators also operational. They should also report those who have an environmental nature where the environment variable has a strategic importance for company. Hence the need to adopt a performance measurement system adequate with the companies needs to have information about the ESCM. To ensure progress for environmental management, top management must be fully committed (Zsidisin and Siferd 2001). Support are needed also from mid-level managers for successful implementation of environmental practices (Bowen et al. 2001). The cooperation and communication between companies departments is important to successful environmental practices (Zhu and Geng 2001; Aspan 2000).

The studies on the performance measurement of the green supply chain and of the ESCM are still relatively few and sparse. At the same time, the principles of accounting are still far to be perceived by their internal environment variable (Bracci and Maran 2013). In the next sections we will analyze the main models of performance management from the perspective of ESCM.

3.1 The Beamon Model (1999)

One of the early works aimed to integrate the environmental variable in the measurement of performance is Beamon (1999). The author highlighted the characteristics of the extended supply chain or green supply chain. In addition to traditional links with various subjects involved (suppliers, manufacturers, distributors, retailers and customers), the author provided other links caused by the set of activities of recycling, reuse and remanufacturing. With the aim of reducing the environmental impacts of products and processes, from the raw material extraction to product disposal, and that allow to the creation of semi-closed loop supply chain. The author argued that it is important, for the design of the extended supply chain and its analysis, to develop appropriate performance measures. The measures have to reflect the dual purpose of the extended supply chain, that is, economic efficiency and environmental protection. So the traditional measures of supply chain performance based on cost, flexibility and quality are not sufficient to evaluate and to respond to companies needs. According to Beamon (1999), a performance measurement systems that respond to the needs of the extended supply chain have to include performance indicators related to different environmental categories. In particular, they should include measures of the quantity of waste generated by the system, measures about the use of energy resources and other natural resources (raw materials and other materials). In addition to the existing operational measures, these measures should be provided with regard to the life cycle of the product and processes. Figure 3 shows the performance measures identified by Beamon (1999) to evaluate the extended supply chain or green.

According to Beamon (1999), it is necessary to carry out a series of activities to implement the extended supply chain (or green). For each product of the supply chain it is necessary to identify all the inputs, outputs, by-products and resources. Then, after developing an adequate system of performance measurement, it is necessary to calculate the composite performance in each phase of the supply chain processes and for each product, assigning a single or matrix of value to the composite performance. This step serves to identify and determine the priorities of intervention between the various processes in order to improve the ecological performance of those with a lower score. From the analysis it should be identified and selected the improvement actions to be undertaken. Beamon (1999) does not specify what mean the supply chain processes. So, the author is unclear whether the identification process and the assignment of a composite measure of performance should also be extended to the production process (or other processes) of suppliers and distributors.

The author does not specify also if the analysis has to stop within the organizational boundary and evaluate processes of plan, source, make, deliver and return that allow to manage the supply chain for that part of activities that are related to the single company. However, with regard to performance measurement, the author has the merit to highlight the need for new measures, these measures have to be mainly related to the consumption and re-use (recovery, recycling, reuse) of the resources in the system and to be centered in details on the products and processes.

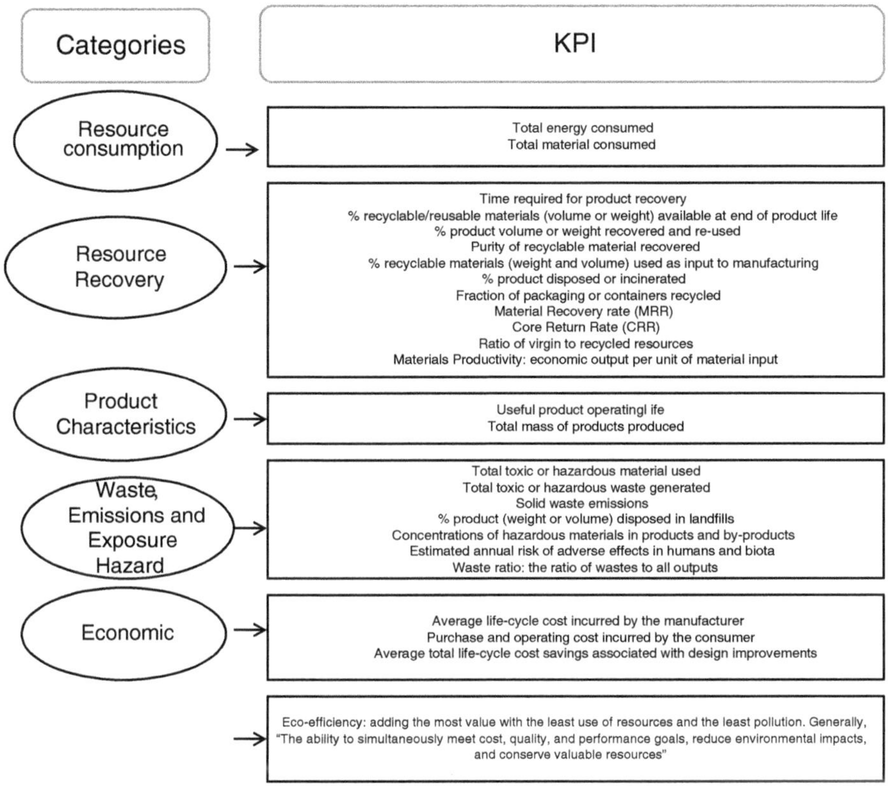

Fig. 3 Extended supply chain performance measures. *Source* Authors' elaboration on data from Beamon (1999)

3.2 The Framework of Hervani et al. (2005)

According to Hervani et al. (2005), performance measurement could be a difficult activity within the organization. When the focus shifts to considering the environmental performance of the supply chain, the performance measurement encounters new difficulties. According to the authors, these difficulties have to be overcome, because the viability of the company and its future competitiveness could depend on the effective adoption of performance measurement of ESCM. The difficulties highlighted by the authors are related to the broad objectives attributed to the system of Environmental supply chain management and performance measurement (ESCM&PM). In fact, according to the authors, every measure must assumed a supply chain perspective. Each entity in the supply chain should be measured and improved according to common goals, all of this requires a very extensive coordination between organizations and creative effort to design new measures.

The evaluation of the performance of supply chain is complicated because there is the need to overcome the distrust with their trading partners, to overcome the difficulties of control variables that are perceived as not to be influenced, to consider different organizations' objectives and measures that some times are not shared between different organizations, to consider the existence of different information systems that not integrated. To develop a performance measurement systems that consider all those variables is necessary to identify effective indicators. These indicators represent the core of the system of ESCM&PM, and allow to evaluate the environmental performance of activities, processes, products and services. The environmental indicators used can be varied and could be related to the emissions, consumption of energy, other environmental impacts, and also include the results of activities for recovery, recycling. A special effort is required for the selection of environmental indicators that measure effectively not only the environmental impacts of operations but also the results of the activities designed to reduce them (suggestions for the choice of indicators can be derived from international standards such as ISO 14001, or from other guidelines such as those of the Global Reporting Initiative (Chiarini 2013).

Hervani and colleagues underlined that *Environmental indicators are plentiful. Yet, there is a difficulty in determining which to use, when to measure them, and how to measure them. Many of these issues must be addressed* (Hervani et al. 2005). These problems may be encountered in the selection of environmental performance indicators with reference to a single company. As the authors reported, the level of complexity is even higher in ESCM PM perspective given the characteristics of inter-organizational and the focus on the life cycle of the product on which it is focused the system. Many of the proposed measures (Table 1) can also

Table 1 Some measures of environmental performance

• Air emission • Underground injection on-site • Release to land on-site • On-site and off-site energy recovery • On-site and off-site recycling • On-site and off-site treatment • Non production release • Source reduction activities • Spill and leak preventions • Raw material modification • Process modification • Product modifications • Pollution prevention opportunity audits • Materials balance audits • Employee and participative management • Management systems pertaining to social and environmental performance	• Number, volume and nature of accidental or non-routine release to land, air and water • Costs associated with environmental compliance • Major awards received • Total energy use • Total electricity use • Total fuel use • Total water use • Total materials use • Habit improvements and damages due to enterprise operations • Quantity of non-product output returned to process or market by recycling and reuse • Major environmental, social and economic impacts associated with life cycle of products and services • Procedures to assist product and service designers to create products and services with reduce adverse life cycle impact

Source Authors' elaboration on data from Hervani et al. (2005)

be used beyond the confines of a single organization with reference to the life cycle of the product, but for this purpose it is necessary to access to data information that must be provided as input by suppliers and customers. It is therefore necessary to reach an agreement on the measures to be used and the identification of sources of data between the main members of the supply chain to create the system. There may be difficulties to be overcome for the realization of a PM system extended to supply chain. This system should be refined gradually, as it is difficult for an organization to develop relationships with their trading partners that allow to immediate access to all the information that necessary their environmental performance. The indicators that will measure the performance of the external environment have be defined in consultation with the most important suppliers, whose activity can generate high impact on the environment and for which there is greater need to carry out a regular monitoring of activities. Such data may be derived from the information systems of the partner or be raised through external audits. Also there is necessity to select the effective indicators for their processes and internal activities to provide a complete picture of the environmental performance of the supply chain.

For the authors, the team responsible for the design of the system of ESCM&PM should not only be inter-functional, that is coming from various internal functions of an organization. But it is preferably to include representatives of suppliers and customers and other stakeholders. In designing the ESCM&PM system it should also consider the environmental management systems of organizations. The system has to be designed and modified according to the principle of continuous improvement based on the Deming wheel (Plan, Do, Check, Act) (Fig. 4) as highlighted by ISO 14031. The team which deals with the design of the system should, then, make sure that this system has the information necessary for an accurate assessment of the environmental performance. To respond to different purposes of external reporting and internal analysis, it may be necessary to modify and refine the system in continuous improvement. The output of ESCM&PM system can serve several purposes, such as external communication, internal analysis or compliance with the regulations. In particular for external communication, companies take the measures that can be compared over time in order to show the improvement of their environmental and the supply chain performance.

The performance measurement system in the ESCM identified by Hervani et al. (2005) focused on environmental issues and in particular on those regarding the supply chain. It can be separated from the other PM systems within the company. In fact, question which the top management have to respond in designing the system is its connection with other internal systems. Then, it could be configured as a model of inter-organizational performance measurement. The level of sophistication required by a system is high, and also the efforts to build and manage it, and the related costs may be particularly high.

This system although is interesting from a theoretical point of view but it might be difficult to be realized from the practical point of view. Mainly due to the difficulty of integrating data from different subjects, but good relations with other subjects involved in the supply chain can help to overcome many of these difficulties. This system is still in early stages, it should to be carried out more research

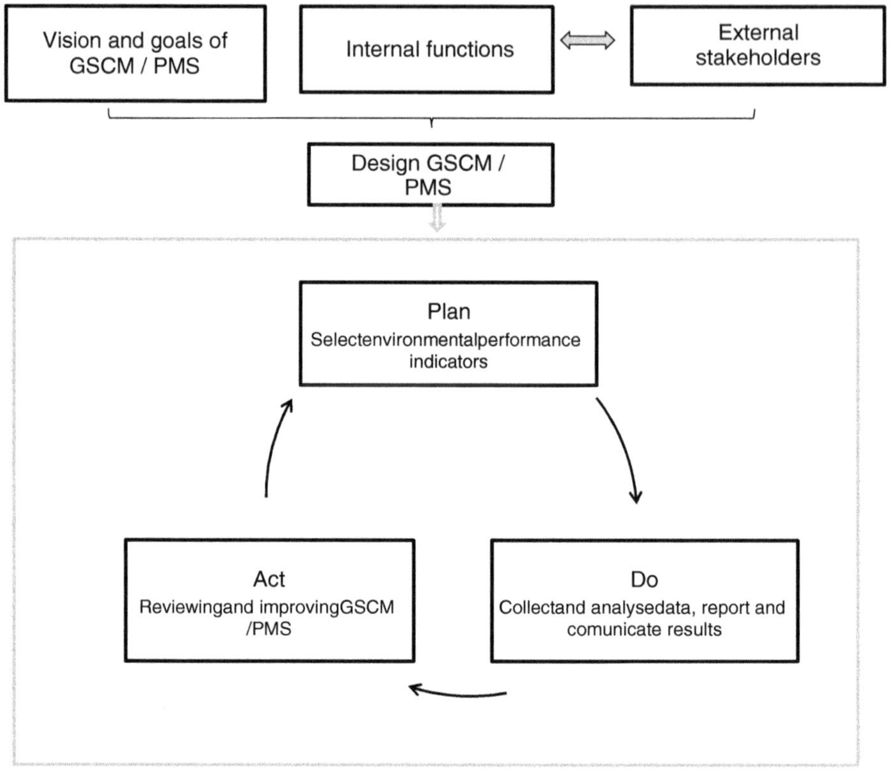

Fig. 4 Performance measurement system and ESCM. *Source* Authors' elaboration on data from Hervani et al. (2005)

to complete their field of study of inter-organizational performance measurement. Measurement systems available may still not be adequate for a comprehensive assessment of ESCM.

3.3 Balanced Scorecard

The model of the Balanced Scorecard (BSC) is suitable for a strategic perspective of control (Kaplan and Norton 1992, 1996). It can be used also to study the relevance to the environmental variable. The original formulation of this model goes back to the early work of Kaplan and Norton (1992). The authors observed that managers in making decisions and in monitoring the performance of the company do not generally rely on a single set of indicators (financial) performance. But they need different types of indicators that relate to different perspectives. So that no one-dimensional system can provide a clear and comprehensive presentation of the results achieved by the organization and focus the attention of managers on all areas and critical dimensions for the future success of the business.

The perspectives may be different and customized for each type of business, those identified by the authors are the economic and financial perspective, customer perspective, internal processes perspective and learning and growth perspective. These perspectives are important for all businesses, which can add other perspectives. The addition of other perspectives is a process that have to be done carefully, because each perspective is linked with other, and because the indicators and objectives have to be coherent. The economic and financial perspective is indicative of the ability of the company to operate according to criteria of cost-effectiveness and achieve adequate profits. The customer perspective is indicative of the organization's ability to satisfy their needs creating value for the same. The internal process is indicative of the ability of the company to be efficient and to have adequate manufacturing processes. Learning and growth is indicative of the ability to develop skills and expertise, new technologies and innovation and is central to the improvement and growth of other prospects in the long term.

The BSC then serves to compose a balanced framework of indicators to guide the action of the manager and allows to control the firm's performance in different perspectives (all equally important to its future success). For Kaplan and Norton the BSC is not limited to be a tool for performance measurement, but it is also a management tool ("a core management system") that should guide the action of the management and guidance to organizational behavior (Kaplan and Norton 1996). The Balanced Scorecard system has to allow the communication of strategic priorities identified, direct the incentive systems (communicating and linking), decline the strategies into a coherent system of sub-goals to determine the good allocation of resources (planning and target setting), and to provide a feedback on the results achieved in all strategic perspectives to guide the process of organizational learning (strategic feedback and learning) (Kaplan and Norton 1996).

Summarized the main characteristics of the model, we can introduce the environmental variable in the discussion. Since the Balanced Scorecard highlights the importance of the strategic use of the indicators of performance in different perspectives (and in doing so it presents itself as a useful tool for the deployment of the strategy). Organizations, adopting a management approach of ESCM, will integrate their strategy considering the objectives related to improving the environmental performance of the organization and its supply chain. The environmental objectives take on strategic importance and are connected in a coherent way with other objectives of the company. Epstein and Wisner (2001) argued that the Balanced Scorecard is an effective tools to translate into action the desire of managers to integrate greater environmental and social sustainability in the strategy of company. The Balanced Scorecard is an appropriate instrument to establish clear links between the resources involved by companies in environmental and social responsibility initiatives and operating results (Epstein and Wisner 2001). For Epstein and Wisner (2001) the Balanced Scorecard can be customized according to the environmental and social objectives of any business, some will have a particular focus on internal processes (and thus may have environmental indicators within this perspective), while others may build environmental performance indicators in all perspectives of the Balanced Scorecard identified by Kaplan and Norton. Other companies could still be able to create a new perspective of the Balanced Scorecard dedicated specifically to the environmental and social dimension. In particular, the adoption of a fifth perspective, according to the authors, is due to the desire to give even greater emphasis to environmental and social objectives, and will be used by those companies that identify in the ability to manage an excellent relationship with the environment the source of possible competitive advantages, or where the environment variable is critical for the type of activity.

In Fig. 5 we list the environmental indicators to be used in different perspectives reported by Epstein and Wisner (2001). The characteristics of these indicators and the objectives to be achieved are also important in the context of ESCM. In the context of ESCM the importance given to the environmental performance of suppliers is very high. Since the aim is to improve the environmental performance not only internal but also at the level of the supply chain, through selection of supplier more environmentally conscious, or through the implementation of environmental projects in collaboration. The BSC may also incorporate a perspective dedicated to suppliers, with the introduction of specific indicators that measure the progress taking place with them. Then, this allows to the evaluation and the subsequent identification of initiatives (of ESCM) to improve the performance in this sense. Examples of these indicators could be the number of suppliers with an environmental management system, the number of projects to be implemented with suppliers to improve environmental performance, the number of recycled packaging from suppliers, the number of production processes reviewed, the number external audits carried out.

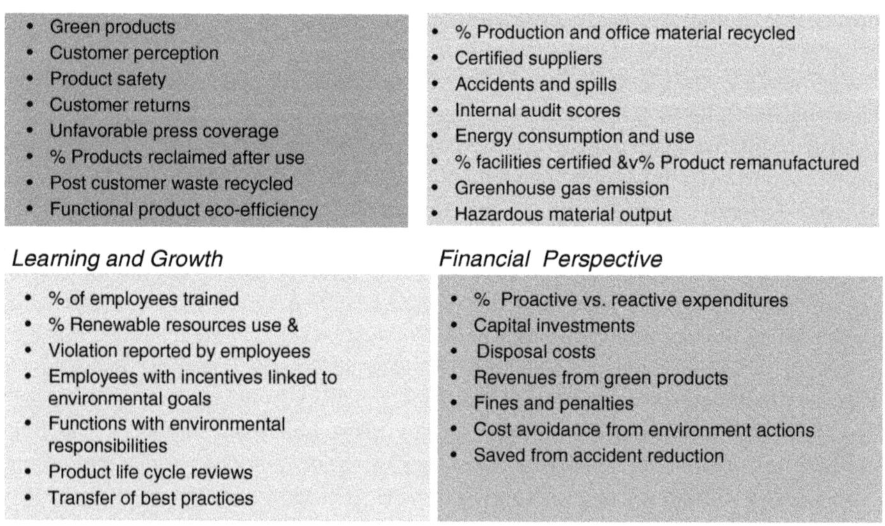

Fig. 5 Example of environmental measures in a Balanced Scorecard. *Source* Authors' elaboration on data from Epstein and Wisner (2001)

The BSC can be an effective tool to deploy the environmental strategy of the company and to recommend the action to be taken to improve the environmental performance of the company and its supply chain. It is therefore a useful strategic tool to support and to implement the initiatives of ESCM. The Balanced Scorecard, as a tool for deployment of the strategy and performance measurement, allows to connect the goal of greater environmental sustainability of company and supply chain management with other business strategies. In doing so, it provides to the management to indicate what are the strategic objectives on which to focus and the indicators to be monitored to assess the degree of their achievement. The managers, after identifying in a clear performance objectives, have to activate the necessary levers, including the practices of ESCM, to achieve its strategic objectives including more sustainable environmental impact of the supply chain.

3.4 Integration, Comparison and Considerations About the Models Proposed

In the previous sections, we introduced the concepts of ESCM and PM and some model proposed in the literature to meld the tow together. We are now going to compare and reflect upon the framework based on the BSC and the Hervani et al. (2005) and by Beamon (1999) model's. Table 2 summarizes the similarity and differences among the models in terms of link with the strategy, internal and external focus and link with other performance dimensions. The table has been drawn considering the commonalities and diversity of the various models analyzed. The table shows the differences between the various models respect to four characteristics: links with the strategy, internal focus, external focus and links with other performance dimensions. While all models give importance to the internal performance (internal focus) in a perspective supply chain, only the model of Hervani (2005) that broadens the scope of measurement also to suppliers. With regard to the relationship with the business strategy, both the balanced scorecard (Epstein and Wisner 2001) and the model of Hervani (2005) have tried to measure the performance of supply chain considering its relationship with business strategy. With regard to the relationship with other dimensions such as the financial performance,

Table 2 The ESCM performance measurement models compared

	Beamon (1999)	Hervani et al. (2005)	Balanced scorecard (Epstein and Wisner 2001)
Links with the strategy	No	Yes	Yes
Internal focus	Yes	Yes	Yes
External focus (i.e. suppliers)	Limited	Yes	Limited
Links with other performance dimensions	Limited	Limited	Yes

and the customers perspective only the balanced scorecard (Epstein and Wisner 2001) that includes other dimensions in the analysis.

The model proposed by Hervani et al. (2005) focused on the integration of environmental management and supply chain strategy. To create a green supply chain it is necessary to adopt a policy of supply chain that also considers the environment and that it is consistent with other corporate strategies. The authors answered to how this supply chain strategy have to be designed and realized. This system is mainly focused on the management of relationships with suppliers, their selection and evaluation. The model proposed by the authors provides guidance on how to implement a green procurement strategy consistent with the other objectives and strategies of organization. The model presented is useful in this context because it allows to control and better manage the environment variable since the input of the company through the management and development of suppliers. Several authors point out that the supply chain management in an environmental perspective, i.e. the creation of ESCM initiatives, should not be in conflict with other goals and other strategies of the organization. As reported by Hervani et al. (2005) the involvement of operational managers, in particular those involved in supply chain management, in the strategic planning process should lead to the formulation of a coherent supply chain strategy with the other strategies of the organization. Also the implementation of the sourcing strategy at the operational level, through the work of cross-functional team, should allow to consider and balance the needs of the various functions, and to a focus on management processes which is fundamental in SCM. Hervani et al. (2005) focused their research on the integration of the environmental management and supply chain strategy. Even if they highlighted the importance of identifying and monitoring appropriate performance indicators, they do not suggest a specific performance measurement system to be used. Since the activity of measuring and monitoring are handled primarily at the operational level by the team responsible of the commodity strategy, the measures for the monitoring and evaluation of the goals set by the specific commodity strategy have to be aggregated to show the results to higher management levels. To this end, the use of other tools such as the BSC could be useful both to support the planning and deployment of environmental strategy on the supply chain. Also it could be useful to show the results arising from the management of the various commodity strategies in aggregated level. The BSC could be a "core management system" useful within the model proposed in the ESCM literature.

As pointed by Epstein and Wisner (2001) the Balanced Scorecard can manage the environment performance, through which the system of performance measurement that focuses on the environmental performance of the supply chain could be created. The authors reported that they are necessary to modify and extend it. However, more research on the tools for measuring performance in inter-organizational perspective are desirable. The BSC was not developed to communicate information on environmental performance, even less at the level of the supply chain. But it can be customized and extended through the choice of key performance indicators relating to the management of the supply chain and different environmental indicators (also related to internal processes) that could be included in the different perspectives identified by Kaplan and Norton perspectives. For example in the "learning and growth" perspective we could include the number of ecological projects undertaken jointly with the main trading partners, the number of analyzes on the environmental impacts associated with the life cycle of the products, research and development in collaboration with the suppliers, the number of hours of training and support provided to partners on environmental issues. For internal management the BSC can be useful tool, which can be integrated with some key measures of performance related to supply chain management and environmental performance. Because it provides information on the results under different analytical perspectives, including economic and financial, which is always important even in the pursuit of reducing environmental impacts. The measures of the BSC should be limited to give a balanced and concise on the achievement of strategic objectives, while at the operational level the measures and information to control and produce should to be more.

The performance measurement system suggested by Hervani et al. (2005) and Beamon (1999) could also be used for external reporting and could have an inter-organizational dimension. In particular the Hervani et al. (2005) model has a specific external focus, in the sense that each performance indicator should have a dimension of supply chain. The system of ESCM PM can be separated from other systems of performance measurement and other corporate information systems, and it could constitute parallel system to those already used by the organization. The performance measurement system can communicate with other managerial processes, provide relevant information to other systems, or used data and information generated by other systems. The architecture and the interface of the system have to be specifically designed by the expert of information systems, according to the objectives and needs of employers and based on the level of collaboration between other subjects of the supply chain. Data and information of the supply chain could be collected by the various operational managers, which also ensure their transmission and sharing, or they could be collected by the company that promote the system. PM systems can gradually rise to its inter-organizational dimension with time and with the increase of commitment in environmental management. Also with the increase of the level of collaboration and integration with other subjects in the supply chain. To provide a system for measuring the performance of such inter-organizational it is necessary to have high level of trust, high integration and collaboration with business partners. The difficulty to manage the system is proportional to the number of subjects involved and therefore depends also on the length and breadth of the supply chain. At the same time, the performance measurement system integrated with the supply chain management may be affected by other operational strategy, such as lean production or six-sigma (Chiarini and Bracci 2013). A system able to generate information about the environmental performance of the supply chain is certainly a useful tool for companies for the purpose of disclosure of their level of commitment in the environment conservation. The information obtained at the level of the supply chain can be communicated together with the other environmental outcomes in the sustainability reports, which has become a useful tool to communicate social activity to the stakeholders.

4 Discussion and Conclusions

We can propose some general reflections on the measurement of performance regarding ESCM. The to ESCM leads companies to reduce the environmental impacts associated with their products in a supply chain perspective. The choice of how to achieve this goal can be different from company to company, it's dependent on the capabilities and available resources. Some organizations will adopt a broader spectrum of activities than others (for example, between those of green design, green procurement, green manufacturing, green distribution and reverse logistics, and within these categories can accomplish different tasks), while some other may seek greater collaboration with suppliers or will have a focus on the control and the imposition of requirements and specifications in their selection and will have other focus mainly internal, even if the creation of strong relationships with other subjects in the SC is crucial to embrace this new management philosophy. Thus, some companies may want to check in detail the environmental impacts of suppliers and their improvement, to collaborate with them in the design of products and processes or new technologies. The focus of the measurement and the evaluation of results in the ESCM system have to be on the environmental performance (but not only), it can be internal or external depending on the circumstances and objectives. For the other subjects involved the measurement and evaluation will be based on the specific relations among them, on the objectives and on the requirements that are required in the implementation of the green supply chain. Some company will be interested to develop an inter-organizational system as suggested by Hervani et al. (2005), while others may decide to control the internal performance, through a BSC for example, and manage suppliers through the application of standards for procurement and environmental certifications of various types.

We can conclude that the system of performance measurement in ESCM have to be coherent with the objectives of the organization looking for a fit with the environment (Melnyk et al. 2014). The extension of this system depends on the degree of involvement and the level of collaboration between all subjects in the supply chain. The collection of data on the environmental impacts of all those involved in the supply chain can be useful mainly for communicating the results of the supply chain to external stakeholders interested in environmental values. This allows to improve its reputation, and to measure progress and its improvement.

More research are desirable to propose more considerations on the structure ad characteristics that systems of PM a supply chain environment are required to have, as well as their effect and uses (Franco-Santos et al. 2007; Melnyk et al. 2014). The research should study and analyze the systems used in practice by those companies that are most committed in the implementation of the ESCM. That organizations have already integrated aspects of environmental management in their strategic business. At present, researches have focused study on performance measurement systems used by companies to manage CSR initiatives, or environmental strategy, less attention given to the ESCM (Arjaliès and Mundy 2013; Rodrigue et al. 2013).

The main contribution of the paper to the extant literature is the attempt to critically analyzing the patchy literature on the performance measurement within an ESCM, proposing the need to look at the two elements in an integrated and connected way. In so doing, three specific models proposed by the literature were analyzed stressing the differences and similarities, but calling for a better understanding of their functioning and use within organization. Further research is thus recommended in order to understand how organization manage their environmental performance along the supply chain and the way it is integrated at the strategy level and performance management systems. Besides, it would be interesting to focus on the intended and unintended consequences in the implementation of performance measurement system to manage the environmental performance along the supply-chain.

References

Arjaliès DL, Mundy J (2013) The use of management control systems to manage CSR strategy: a levers of control perspective. Manag Acc Res 24(4):284–300

Aspan H (2000) Running in nonconcentric circles: why environmental management isn't being integrated into business management. Environ Qual Manage 9(4):9–75

Atkinson AA, Waterhouse JH, Wells RB (1997) A stakeholder approach to strategic performance measurement. Sloan Manag Rev 38:25–38

Azevedo S, Carvalho H, Cruz-Machado V (2013) Using interpretive structural modeling to identify and rank performance measures: an application in the automotive supply chain. Baltic J Manag 8(2):208–230

Beamon BM (1999) Designing the green supply chain. Logistics Inf Manag 12(4):332–342

Bititci U, Garengo P, Dorfler V, Nudurupati S (2012) Performance measurement: challenges for tomorrow. Int J Manag Rev 14(3):305–327

Bowen FE, Cousins PD, Lamming RC, Faruk AC (2001) Horse for courses: explaining the gap between the theory and practice of green supply. Green Manag Int Autumn 41–60

Bratic D (2011) Achieving a competitive advantage by SCM. IBIMA Business Review

Bracci E, Maran M (2013) Environmental management and regulation: pitfalls of environmental accounting? Manag Environ Qual Int J 24(4):538–554

Carter RC, Carter JR (1998) Interorganizational determinants of environmental purchasing: initial evidence from the consumer products industry. Decis Sci 29:28–38

Chiarini A (2013) Designing an environmental sustainable supply chain through ISO 14001 standard. Manag Environ Qual Int J 24(1):16–33

Chiarini A, Bracci E (2013) Implementing Lean Six Sigma in healthcare: issues from Italy. Publ Money Manag 33(5):361–368

Epstein M, Wisner P (2001) Good neighbours: implementing social and environmental strategies with the BSC. Balanced Scorecard Rep 3(3):8–11

Femenias LB, Llach J, Almeida MMA (2013) Is the adoption of environmental practices a strategical decision for small service companies? Manag Decis 51(1):41–62

Franco-Santos M, Kennerley M, Micheli P, Martinez V, Mason S, Marr B,Gray D, Neely A (2007) Towards a definition of a business performance measurement system. Int J Oper Prod Manage 27(8):784–801

Green KW, Zelbst PJ, Meacham J, Bhadauria VS (2012) Green supply chain management practices: impact on performance. Supply Chain Manag Int J 17(3):290–305

Gunasekaran A, Patel C, Tirtiroglu E (2001) Performance measures and metrics in a supply chain environment. Int J Oper Prod Manag 21(1/2):71–87

Gunasekaran A, Patel C, McGaughey RE (2004) A framework for supply chain performance measurement. Int J Prod Econ 87(3):333–347
Handfield R, Sroufe R, Walton S (2005) Integrating environmental management and supply chain strategies. Bus Strategy Environ 14(1):1–19
Hervani AA, Helms MM, Sarkis J (2005) Performance measurement for green supply chain management. Benchmarking Int J 12(4):330–353
Huo B, Qi Y, Wang Z, Zhao Z (2014) The impact of supply chain integration on firm performance. Supply Chain Manag Int J 19(4):369–384
Kaplan RS, Norton DP (1992) The balanced scorecard—measures that drive performance. Harvard Bus Rev 70(1):71–79
Kaplan RS, Norton DP (1996) Linking balanced scorecard to strategy. Harvard Bus Rev 39(1):53–79
Liu X, Wang L, Dong Y, Yang J, Bao C (2011) Case studies of green supply chain management in China. Int J Econ Manage Eng 1(1):22–34
Melnyk MA, Bititci U, Platts K, Tobias J, Andersen B (2014) Is performance measurement and management fit for the future? Manag Acc Res 25(2):173–186
Mentzer JT, DeWitt W, Keebler JS, Min S, Nix NW, Smith CD, Zacharia ZG (2001) Defining supply chain management. J Bus Logistics 22(2):1–25
Ninlawan C, Seksan P, Tossapol K, Pilada W (2010) The implementation of green supply chain management practices in electronics industry. In: Proceedings of the international multiconference of engineers and computer scientists, vol 3
Perera S, Harrison G, Poole M (1997) Customer-focused manufacturing strategy and the use of operations based non-financial performance measures: a research note. Acc Organ Soc 22 (6):557–572
Rao P, Holt D (2005) Do green supply chains lead to competitiveness and economic performance? Int J Oper Prod Manag 25(9):898–916
Rodrigue M, Magnan M, Boulianne E (2013) Stakeholders' influence on environmental strategy and performance indicators: a managerial perspective. Manag Acc Res 24(4):301–316
Seuring S, Müller M (2008) From a literature review to a conceptual framework for sustainable supply chain management. J Clean Prod 16(15):1699–1710
Simon R (1995) Levers of control: how managers use innovative control systems to drive strategic renewal. Harvard Business School Press, Boston
Srivastava SK (2007) Green supply-chain management: a state-of-the-art literature review. Int J Manag Rev 9(1):53–80
Sun SY, Hsu MH, Hwang WJ (2009) The impact of alignment between supply chain strategy and environmental uncertainty on SCM performance. Supply Chain Manag Int J 14(3):201–212
Testa F, Iraldo F (2010) Shadows and lights of GSCM (Green Supply Chain Management): determinants and effects of these practices based on a multi-national study. J Clean Prod 18 (10):953–962
Thun JH, Müller A (2010) An empirical analysis of green supply chain management in the German automotive industry. Bus Strategy Environ 19(2):119–132
Vachon S, Klassen RD (2006) Green project partnership in the supply chain: the case of the package printing industry. J Clean Prod 14(6–7):661–671
Zhu Q, Geng Y (2001) Integrating environmental issues into supplier selection and management: a study of large and medium-sized state-owned manufacturers in China. Green Management International, Autumn 27–40
Zsidisin GA, Siferd SP (2001) Environmental purchasing: a framework for theory development. Eur J Purchasing Supply Manag 7(1):61–73
Zhu Q, Sarkis J (2004) Relationships between operational practices and performance among early adopters of green supply chain management practices in Chinese manufacturing enterprises. J Oper Manag 22(3):265–289
Zhu Q, Sarkis J, Lai KH (2008) Green supply chain management implications for "closing the loop". Transp Res Part E Logistics Transp Rev 44(1):1–18

The Measurement of Environmental Performance in Hospitals: A Systematic Review of Literature

Andreia Pasqualini Blass, Sergio E. Gouvea da Costa, Edson Pinheiro de Lima and Lilian Adriana Borges

Abstract For the past 25 years, academics and practitioners have raised concerns on the measurement of environmental sustainability performance. Addressing these issues has become paramount for organizations worldwide. In the healthcare sector, especially in hospitals, measuring environmental performance relates to the reduction of environmental impact and continuous improvements in the quality of processes and outcomes. Considering this context, the study seeks to identify how concerns on environmental sustainability and performance measurement have pervaded the healthcare operations management literature. A two-tier systematic review of literature included: (1) a literature review on performance measurement, environmental sustainability and healthcare operations management; (2) a bibliometric review of literature, which evaluated published studies (1988–2013), focusing on citation numbers, country of origin, main journals, authors and themes. The study highlighted concerns on the lack of strategic focus of performance indicators, relevance and robustness of metrics and difficulties for the deployment of measures within different hierarchical levels. Furthermore, the bibliometric review emphasized the scarcity of published research addressing the environmental performance measurement in the healthcare setting, especially in the context of developing countries. New frameworks are necessary to define relevant and meaningful indicators for monitoring and assessing environmental performance if healthcare systems and operations are to be improved.

A.P. Blass (✉) · S.E.G. da Costa · E.P. de Lima
Pontifical Catholic University of Parana—PUCPR, Curitiba, PR, Brazil
e-mail: apasqualini@unidavi.edu.br

S.E.G. da Costa
e-mail: s.gouvea@pucpr.br

E.P. de Lima
e-mail: e.pinheiro@pucpr.br

S.E.G. da Costa · E.P. de Lima
Federal University of Technology, Curitiba, PR, Brazil

A.P. Blass · L.A. Borges
UNIDAVI, Rio Do Sul, SC, Brazil
e-mail: lilian.borges@unidavi.edu.br

A. Chiarini (ed.), *Sustainable Operations Management*,
Measuring Operations Performance, DOI 10.1007/978-3-319-14002-5_5

Keywords Healthcare · Operations management · Environmental performance · Hospitals

1 Introduction

For the last 25 years, organizations have been urged to internalize concerns related to the use of their productive resources and how to rationalize the impact of their processes on the environment (Kleindorfer et al. 2005). Environmental sustainability[1] has become an important competitive factor that influences strategies, demanding the adoption of 'best practices'. Developing robust performance measurement frameworks to evaluate environmental sustainability has become a continuous learning process. Sustainable development indicators translate sustainability issues into quantifiable environmental performance measures. Using meaningful indicators to evaluate performance allows monitoring and reflection on the measurement process (Bourne et al. 2000), since their main goal is to assist with the search and achievement of superior performance (Scerri 2010).

According to Porter (2010), enhanced improved performance in any sector depends on the existence of a common goal for the activities and interests of different stakeholders. In the healthcare sector, this goal is associated with the definition of value represented by the patients' health outcomes by the unit of currency invested in the process. If this value is improved, patients, managers, service providers and suppliers can benefit from it while the sustainability of the healthcare system is also improved (Porter 2010).

Performance measures also encompass concerns on the quality of healthcare services (Kanji and Sá 2003). Quality improvements may lead to less waste of resources, improved patient satisfaction and medical care effectiveness (Maki et al. 2008). In the past, the incineration of waste in local hospitals was common practice (Department of Environment (DoE) 1990). New regulations have forced these organizations to solve this problem through a more systematic and selective manner (Tudor et al. 2005) to promote environmental sustainability. Nowadays, healthcare services providers should consider an efficient mobilization of resources in their blueprints (Vilaça and Oliveira 2008). A healthcare system includes activities and structures, whose primary purpose is influencing health. The health of the population is determined by several factors and among these factors is the provision of healthcare services. Healthcare organizations that 'embrace' environmental sustainability tend to gain significant performance improvements and decrease costs related to waste disposal (Jarousse 2012). Efforts to improve sustainability indicators may also raise awareness and generate consistent attitudes to preserve resources.

[1] The terms environmental and sustainability are used interchangeably and in combination in the study.

Healthcare indicators correspond to measurable and explicitly defined items related to the structures, processes and outcomes of the provision of services (Campbell et al. 2001). They represent statistical instruments that monitor performance, analyzing the resources deployment within these systems without necessarily judging value on quality.

The combination of the internalization of concerns on environmental sustainability, the importance of an efficient management of resources and the improvement in service quality raise questions such as:

- How have the concerns on environmental sustainability pervaded the healthcare operations management?
- What are the main environmental performance measurement frameworks used within healthcare organizations?
- What are the main themes being discussed in the context of healthcare organizations?

The study seeks these answers through a systematic review of literature. The paper is divided in five main sections. Section 2 discusses the research methods and strategy adopted in the study. Section 3 analyzes the application of performance measurement frameworks and metrics/indicators used to address environmental sustainability within the healthcare context. The Brazilian legal framework related to healthcare and current practices are also briefly evaluated. Section 4 presents the results of the bibliometric review of literature. Section 5 summarizes final considerations on the study findings while approaching the next steps of the research.

2 Research Methods

Initially, a literature review on performance measurement, environmental sustainability and healthcare operations management was conducted. The main objective was to identify essential terms that convey knowledge across published studies (Krauthammer and Nenadic 2004). Moreover, the literature review indicates what has been learned about the topics and where research gaps are present (Webster and Watson 2002). After this initial step, the bibliometric technique was chosen as research strategy. Bibliometric methods indicate research impact (Wallin 2005; Van Raan 1996; Van Raan and Van Leeuwen 2002). Four databases were selected for the search: Web of Science®, Medline Complete® (EBSCO), Wiley Online Library® and Scopus® (Elsevier). Web of Science was chosen due its multidisciplinary and comprehensive nature. It includes Science Citation Index Expanded (SCI-EXPANDED), Social Sciences Citation Index (SSCI), Arts & Humanities Citation Index (A&HCI), Conference Proceedings Citation Index-Science (CPCI-S) and Conference Proceedings Citation Index—Social Science and Humanities (CPCI-SSH). Medline complete was selected for two reasons: (a) focus on Health Sciences/Biomedical publications that cover healthcare operations; (b) consistent reputation by being associated with the US National Library of Medicine.

Wiley Online was chosen mainly because of its comprehensiveness; it is regarded as the world's 'broadest multidisciplinary resources'. Finally, Scopus was selected because of the emphasis placed on the peer-review process of its resources and the inclusion of articles in press and conference proceedings.

The searches were conducted between January and March of 2013. The combination of terms found in the literature review guided these searches. The following terms were combined: 'Hospital', 'Operations', 'Environment*' (the asterisk symbol was used to obtain results that included environment and environmental), 'Indicators', 'Performance', 'Sustain*' (the asterisk symbol was used to obtain results that included sustainable and sustainability), 'Framework', 'Quality', 'Health' and 'Care' (the terms were combined in the searches, since both Healthcare and Health Care are found in literature). Papers/articles published between 1988 and 2013 were included (25 years), since the publication of 'Our Common Future' (1987) represents an important milestone that defined terms currently used to approach concerns on the environment and raise awareness on the importance of the issue (Atkinson 2000; Kleindorfer et al. 2005; Linton et al. 2007). The combinations of terms were used as topics to be found in the Abstracts of the articles.

In order to avoid researcher bias, two independent searches took place: one by the first author and another by one of the co-authors. The coding of different themes followed the rationale: Title of the article→Abstract→Keywords. To avoid bias, the coding process followed the same strategy with two independent analyses. Similar results were obtained in both cases.

3 The Measurement of Environmental Performance in Healthcare Organizations

3.1 *Performance Measurement and Environmental Sustainability*

The first step for the measurement process refers to the conception of performance measures. According to Medori and Steeple (2000), Gomes et al. (2004) and Neely (2004), several frameworks have been proposed in literature:

- The performance measurement matrix (Keegan et al. 1989);
- The SMART model (Cross and Lynch 1989);
- Performance measurement for World Class Manufacturing created by Maskell (1989);
- The performance measurement questionnaire (Dixon et al. 1990);
- The framework proposed by Beischel and Smith (1991);
- The model based in services of Fitzgerald et al. (1991);
- The Balanced Scorecard (Kaplan and Norton 1992);
- The Business Excellence Model developed by the European Foundation for Quality Management (EFQM 2013);

- The model of organizational macro processes proposed by Brown (1996);
- The dynamic integrated performance measurement system (Ghalayini et al. 1997);
- The performance measurement record sheet (Neely et al. 1997); and
- The performance PRISM (Neely et al. 2002).

Even though a number of performance measurement frameworks can be found, a gap between the application and validation of the concepts within these models has been observed (Pinheiro De Lima et al. 2013). Gomes et al. (2004) found, after an extensive review of literature on performance measurement, little evidence that these measures are formalized and connected with organizational strategy and efficiency. Kaplan and Norton (2004) highlight the need for a correlation between strategy and performance measurement. Only through a consistent connection among these elements, strategies can be translated into actions to achieve strategic goals (Kaplan and Norton 2004; Otley 1999; Wongrassamee et al. 2003) could not find any specification as how to formulate these goals within known models.

Feng and Joung (2011) searched for an appropriate model to evaluate the three dimensions of sustainability in manufacturing organizations: economic, social and environmental. In their study, Feng and Joung (2011) found that the focus tends to be on the perception of external stakeholders (reporting) rather than in getting required information for the decision-making to improve performance. Table 1 presents some of these models. The information contained in Table 1 was complemented with additional frameworks reviewed by the authors of the present paper.

Table 1 Frameworks/models for sustainable performance measurement

Frameworks	Number of indicators
Corporate Sustainability Index (ISE)	14 criteria
Dow Jones Sustainability Index (DJSI)	Index based on 12 criteria
Eco-indicators 1999	Indicator based on three factors
Environment Performance Index (EPI)	19 indices
Environmental Sustainability Index (ESI)	68 indicators
European Union Environmental Pressure Indicators (EUEPI)	60 indicators
Ford Product Sustainability Index (FPSI)	8 indicators
Global Reporting Initiative (GRI)	91 indicators
GM Metrics for Sustainable Manufacturing	30 indicators
ISO 14031 Environmental Performance Evaluation	155 indicators (examples)
OECD Core Environmental Indicators	46 indicators
The Earth Charter	16 principles
The ETHOS Program	7 dimensions
The Global Compact	10 principles
The Millennium Development Goals	8 objectives
UNCSD Indicators of Sustainable Development	96 indicators
Walmart Sustainable Product Index (WSPI)	15 questions

In addition to the models mentioned in Table 1, researchers have suggested the combination of existing frameworks. For instance, Nikolaou and Tsalis (2013) combined the Balanced Scorecard rationale with the indicators proposed within the Global Reporting Initiative (GRI). Nonetheless, after reviewing the frameworks, some difficulties reported in literature can be cited:

- Deployment of performance measures: studies on sustainable performance measurement reveal difficulties in considering context-specific features in terms of the definition of indicators and deployment of performance measures for each hierarchical level within the organization;
- Prevalence of subjective judgements: in general, the delineation of performance measures is associated with a significant subjective component. Measures that are useful for one type of organization may be superfluous for another;
- Scoring methods: Feng and Joung (2011) highlighted the excessive simplicity of scoring methods proposed in the models. Essential aspects related to corporate strategy, changes that may occur in the strategic path of a company/business unit and required trade-offs are consistently disregarded.

3.2 Healthcare Operations and Performance Measurement

Bertrand and De Vries (2005) indicate that a number of characteristics differentiate healthcare operations from manufacturing settings. Healthcare operations are regarded as complex and employing performance evaluation techniques used in manufacturing is described as troublesome. Bertrand and De Vries (2005) highlight the delicate balance of power among different stakeholders: managers, medical experts and nurses. Each group has different expectations on organizational performance goals. Service providers are highly trained professionals, who give instructions and received instructions from final consumers (patients). Healthcare services are not commodities that can be stored; they are associated with the management of limited resources. As healthcare organizations are included in more complex healthcare systems containing hospitals, private clinics, hospices and carers, the performance measurement task is also magnified (Curtright et al. 2000). Table 2 summarizes some of these differences.

Table 2 Comparison between manufacturing and healthcare operations (Bertrand and De Vries 2005)

Characteristics	Manufacturing	Healthcare
Object	Materials flow	Patients flow
Final product specification	Previously specified	Subjective and vague
Means of production	Equipment/staff	Equipment/staff
Protection	Inventory/lead time	Waiting Time/lead time
Financial goal	Profit	Control of costs
Market environment	Market competition	Limited market competition

Yearly, the United Kingdom National Health Service (NHS) produces 600.000 tons of clinical, pharmaceutical, infectious and domestic waste at a cost of 42 million pounds (approx. 64 million dollars) (Coote 2002). The healthcare sector has been urged to demonstrate efficiency and effectiveness in service provision (Weir et al. 2009). Healthcare (hospitals, private clinics, etc.) is an open system, which is subject to several influences: social, economic, political and technological that can benefit or negatively affect their services (Oliveira 2003). People undergoing treatment, materials and equipment, specialized healthcare services and supporting activities such as laundry, maintenance, Information Technologies represent the inputs of this system. The process is composed by the diagnosis and treatment of patients, that is, activities performed for the provision of medical services. Treated/cured people, the knowledge acquired through research and process wastes represent the outputs (Oliveira 2003). In the healthcare sector, the provision of public and private healthcare services coexists. Healthcare service providers represent 'the intermediate' between those who finance the healthcare system and those who pay and use these services (Swayne et al. 2006). These organizations are responsible for managing resources on behalf of those who utilize their services. Nonetheless, as healthcare operations evolve, much attention has been paid to the 'healthcare business' instead of focusing on quality improvements (Swayne et al. 2006). Growing pressures for cost reduction and quality improvement have been associated with stricter rules and regulations. This has forced healthcare professionals and managers to reexamine the way performance is evaluated (Castañeda-Méndez et al. 1998).

3.3 *Difficulties in Measuring Sustainability—Metrics and Indicators*

In spite of the efforts related to obtaining environmental sustainability in the healthcare sector, the outcomes of these initiatives have been unsatisfactory (Phillips et al. 2002). Volumes of waste generated in the healthcare have progressively risen (Woolridge et al. 2005). Several policy weaknesses have been highlighted. Tudor et al. (2008) conclude that a more effective participation in the recycling and waste management in the healthcare sector can be achieved through the incorporation of sustainability concepts into the organizational policies; a comprehensive communication plan; training programs and professional development; and emphasis on the benefits from sustainable practices.

A profusion of frameworks have tried to overcome the complexity of the performance measurement process. Difficulties related to the quantification of information and disregard for performance measurement systems already in place have been reported by the literature. A multiple stakeholder perspective is indicated as one of the most important characteristics that contribute to the success of the application of performance evaluation models given the complexity of the healthcare operations (Tawfik-Shukor et al. 2007).

In addition to existing frameworks, other tools have been proposed for the evaluation of environmental sustainability. Paju et al. (2010) suggested the use of the Sustainable Manufacturing Mapping (SMM), which is a combination of Value Stream Mapping (VSM), Discrete Event Simulation (DES) and Life Cycle Assessment (LCA). The US Environmental Protection Agency has introduced a set of evaluation tools that aimed to improve the environmental performance of American companies (Environmental Protection Agency 2007). Hardi and Pinter (1995) show that indicators that aggregate heterogeneous dimensions are not readily available to be considered in the decision-making process. Indicators currently used by organizations aggregate dimensions that are not comparable and, when they are, the evaluation based on the perception of stakeholders should be careful. Moreover, a process for the selection of indicators focused on specific issues may not compose a balanced view of sustainability issues (Hardi and Pinter 1995). Organizational decision processes guided by a reduced set of indicators may require substantial structural changes. This restrictive process may lead to the selection of indicators related to very specific, less meaningful and non-representative targets. Performance indicators should have specific characteristics to measure performance and represent a source for benchmarking and strategy planning (Feng and Joung 2011). Such characteristics include:

- Measurability: an indicator should be capable of measuring quantitatively or qualitatively multidimensional perspectives;
- Relevance: an indicator should be present an useful meaning for the evaluation processes;
- Clarity: an indicator should be easily understood by the community, especially those that are not experts;
- Reliability: information proposed by an indicator should be reliable and useful;
- Data accessibility: an indicator should be based on accessible data;
- Opportunity: the measurement of an indicator should happen with the frequency that allows an informed and well-timed decision making process;
- Long term view: an indicator should be compatible with an open pattern that supports the need for recorded information for future generations.

Moreover, the definition of performance indicators depends on the answers to three fundamental questions (Weir et al. 2009):

- Who should be involved in the development of performance indicators?
- Who is the audience for the performance indicators (stakeholders)?
- What level of performance should be measured?

For instance, despite the wide scope of indicators proposed in the applications of the BSC in healthcare organizations, environmental performance indicators are not usually declared. When environmental measures are present, they tend to refer to existing regulations and/or sustainability programs. These concerns are then associated with the framework with no changes/adaptations. In specific cases in Brazil, some environmental indicators were proposed as reported by Campos and Selig (2002):

- Number of complaints related to the environmental aspect and placed by the community (NGOs, neighborhood, supplier, third-parties, etc.);
- Number of complaints related to any environmental factors made by a visitor divided by the number of visitors per days (percentage);
- Water consumption per person (monthly water consumption divided by the number of people, including visitors, employees and third-parties);
- Electricity consumption per person (monthly consumption divided by the number of people per month, including visitors, employees and third-parties);
- All legal parameters of discharge of effluents required by legislation;
- Percentage of inert waste sent to landfills (tons of inert/generated waste);
- Number of legal non-conformities registered per year (fines, contaminations, violations).

Metrics are vital elements to measure performance, since potential difficulties can be indicated by the gap metric-standard (Melnyk et al. 2004), especially regarding issues related to sustainability (Gunasekaran et al. 2001). Nonetheless, the existence of a framework that deals with the deployment of measures and metrics in strategic, tactical and operational levels to include tangible, intangible, financial and non-financial aspects is questioned in the literature (Gunasekaran and Spalanzani 2012). Metrics mentioned by Zhu and Sarkis (2004) include reductions of air emissions and waste of water, increased investments in training and 'green' purchases, and reduced costs with waste treatment and energy consumption.

Two perspectives should be present to define performance measures: resources and value (Fiksel et al. 1999). Resources are represented by energy, materials, water, land, waste, cost and human capital. Value corresponds to the functional performance of resources such as information content, client satisfaction, environmental quality, business competence, human health and social well-being. BP Amoco™, for instance, uses the production process water discharge as a resource indicator. Volvo™ evaluates how many managers include lifecycle tools in the product development process (value).

Townend and Cheeseman (2009) proposed detailed recommendations on how to measurement environmental performance in healthcare organizations based on (a) general management; (b) social issues; (c) health and safety; (d) energy and water use; (e) purchasing and supply; (f) waste management (responsibility, segregation, storage and packaging); (g) waste transport; (h) recycling and re-use; (i) waste treatment; and (j) final disposal.

Veleva et al. (2003) used the hierarchy of indicators established by the Lowell Center of the University of Massachusetts in their study in the pharmaceutical industry. Level 1 indicators are managed by voluntary initiatives or based on regulations (ISO and others). The eco-efficiency measures (Level 2) translate the connection between the environmental performance and savings of financial resources. The standards and methods for the definition of Level 3 indicators (environmental impact) have been developed only recently. Organizations tend to avoid establishing connections between the indicators and competitive advantage, according to the authors. The reduced influence of the government and stakeholders in a wider adoption of these indicators was highlighted. Levels 4 and 5 refer, respectively, to the availability of data on the supply chain participants and collaborations with entities outside the organization (Veleva et al. 2003).

Kaplan and Norton (1992) developed the Balanced Scorecard, using four perspectives to measure performance: (a) financial, (b) client, (c) internal processes and (d) learning and growth. Gurd and Gao (2008) studied the adaptation of the Balanced Scorecard indicators to measure performance in hospitals and clinics. Some examples can be cited:

- Financial perspective: profit growth indicators, productivity indicators;
- Client perspective: patient retention, patient acquisition, patient satisfaction indicators;
- Internal processes perspective: patient satisfaction, safety and health, productivity, innovation indicators;
- Learning and growth perspective: human capital, information capital, continuous innovation, organizational capital.

However, considering 22 cases reported in their study, only 77 % included the client or patient perspective. Health does not seem the central focus of these BSC applications, the researchers concluded. In the healthcare sector in the US, Mannion and Davies (2002) recognize the existence of scorecards, service provider profile evaluation and medical professional profile. A sufficient number of metrics represents another concern in the performance measurement in hospitals. Multiple aspects are present in a hospital and trying to measure every single aspect of a complex system may emphasize metrics instead of concerns on the quality of the patients and system (Forrest et al. 2006). Regarding environmental performance measurement in healthcare, it is important to point out the existence of common indicators from either practice or current legislation. Table 3 presents some of these common indicators.

Table 3 Performance measurement frameworks in healthcare—common indicators

Indicators related to	Campos and Selig (2002)	Townend et al. (2009)	Veleva et al. (2003)	Gurd and Gao (2008)
Legal nonconformities (fines, notifications)	+	+	+	−
Complaints from the community and clients/patients	+	+	−	+
Water use/consumption	+	+	+	−
Energy use/consumption	+	+	+	−
Gas emissions	+	+	+	−
Use of renewable energy sources/biodiversity	−	+	+	−
Heavy metal emissions (water)	−	+	+	−

(+) Present; (−) Absent

McGlynn (Paju et al. 2010) discusses six actions that are necessary to measure performance in healthcare:

1. Identify the perspectives of the system stakeholders;
2. Develop a framework of responsibilities;
3. Establish explicit criteria on how the system will be evaluated;
4. Select a subset of indicators for periodical reports;
5. Minimize conflicts between financial and non-financial incentives and quality goals in healthcare;
6. Facilitate the development of information systems to support quality monitoring.

Indicators should satisfy one or more stakeholder groups (McGlynn 1997). Learning is a dimension needed in the performance measurement process (Wilcock and Campion-Smith 1998). Scientific soundness and applicability should be achieved. It relates to three aspects: reliability, validity and adjustability. Reliability means that replication of the measurement will produce the same results. Validity refers to the measurement of quality related to healthcare. And adjustability involves the use of other factors apart from quality that will influence the measurement final result. Applicability refers to a consistent choice by managers, covering a number of areas and services. It involves parsimony and comprehensiveness (Wilcock and Campion-Smith 1998).

Although a profusion of models and indicators to measure environmental performance have been proposed in literature, initiatives related to the environmental sustainability tend to lack strategic focus. The main purpose behind these initiatives has been connected with rules and regulations, accreditation requirements and elements pertaining to quality programs (e.g. ISO, The Business Excellence Model). As a consequence of the lack of strategic focus and use of standard measures, the deployment of measures in different hierarchical levels has become troublesome (Fiksel et al. 1999). In developing countries such as Brazil, healthcare operations face major challenges such as the high cost of healthcare services, limited existence of productive resources, inadequate infrastructure/structure (e.g. access to clean water, electricity), lack of public investments and strategically focused governmental policies, which underline the difficulties associated with the measurement of sustainability in healthcare operations.

3.4 The Brazilian Regulatory Framework

In Brazil, nine normative documents deal directly or indirectly with the control and disposal of hospital waste. The evaluation of organizations that provide healthcare services, NA2 Rev. 01 of 6 March 2006, is necessary to acquire Hospital Accreditation with the National Organization for Accreditation (ONA, abbreviation in Brazilian Portuguese). The National Environment Council (CONAMA) and the National Health Surveillance Agency (ANVISA) have specific resolutions. While the Brazilian Technical Standards Association (ABNT) issued six standards on hospital waste. The CONAMA Resolution nº 358 (29 April 2005) approaches the need to treat and dispose of waste from healthcare services. The ANVISA Resolution nº 306 (7 December 2004) addresses the management of waste from these services. The ABNT legislates on:

- Waste transport (NBR 14652, 11 July 2013);
- Methods for the collection of perforating/cutting materials (NBR 13853, de 30 June 1997);
- Internal and external collection of waste under hygienic and safe conditions (NBR 12810, 1 April 1993);
- Classification of materials and definition of terms (NBR 12808, 1 April 1993, and NBR 12807 of 15 June 2013, respectively);
- Required procedures for the intra-establishment management of healthcare services waste (NBR 12809 of 19 May 2013).

Despite the existence of a legal framework, Da Silva et al. (2005) concluded from a study on 91 healthcare facilities, including hospitals (2), health centers (48) and clinical laboratories (22) that in Brazil the practices in most cases do not comply with current legislation. According to Nazar et al. (2005), the mere existence of regulations does not guarantee the success of environmental initiatives. In Brazil, a scarcity of regulations that define the consumption of other resources such as water, electricity and fuel is noteworthy. Recently, some initiatives have tried to address the disposal of liquid waste (part of the solid waste category) by creating Eco-centers and Effluents Treatment Facilities (Vieira and Rodrigues 2013). Concerns on measurement environmental performance are regarded as insufficient and limited due to a number of issues (La Forgia and Couttolenc 2008).

3.5 Environmental Performance Measurement Frameworks Used in Brazil

The United Kingdom public healthcare system (NHS) represents one of the most recognizable performance measurement frameworks described in literature. In the case of the NHS, a national model of performance evaluation has been applied. A set of 51 indicators (2002) is divided and classified in the following dimensions: health improvement, fair access, effective provision of health, efficiency, patient/carer experience and health outcomes (Woolridge et al. 2005). The environmental sustainability indicators, in this case, are designed from current government policies. The EPA 90 (Environmental Protection Act) constitutes the integrative program, which guides these initiatives. Local initiatives are also implemented to promote a sustainable behavior. These programs had a central objective to improve waste management, establishing goals and indicators for recycling and recovery. Clinical waste was classified and a number of strategies were developed to motivate the separation of items. Alternative treatment technologies were proposed whereas risks for the health of human beings had been observed (Tudor 2007). Potential benefits from the separation and recycling of waste have been emphasized by hospitals and other institutions within the British healthcare system. This promoted new initiatives for efficiency and effectiveness improvements (Tudor 2007).

Regarding solid waste, indicators such as Kilograms per solid waste by patient bed/day or Kilograms of solid waste per appointment and per day have been suggested (Oliveira 2003). Indicators are obtained through sampling. There is an apparent consensus found in literature that it is impossible to define a single indicator of sustainable development. A substantial number of indicators are necessary to capture all the important aspects of a particular system (Becker 1997; Bossel 1999). Performance indicators should perform as an integrated information network and should not be analyzed individually. There are knots that connect or relate to information and data that can assist in the analysis of a particular system (Becker 1997; Bossel 1999). These knots are represented by the performance indicators (factors) built from a referential model proposed for the analysis. Those with greater importance (more weight/importance according to stakeholders) can be prioritized in the management actions (Silva et al. 2008). Performance indicators raise awareness and understanding on issues; inform the decision making process; and measure the achievement of established goals (Ventura et al. 2010).

The Syrian Lebanese Hospital (HSL) case represents an environmental performance measurement framework based on quality programs. Launched in 1998, the hospital's environmental management program includes the hospital hygiene service, selective collection of waste and different maintenance solutions. Its main goal is to save resources such as water, electricity and gas. Since its launching, the program generated yearly savings of one million cubic meters of water and 679 tons of paper; 19 tons of glass, 23 tons of aluminum and 101 tons of plastic have been recycled (Bio2 Sustentabilidade 2012). In the hospital, the environmental management project focuses on two primary objectives: a better utilization of input material such as water, electricity and gas and the management of waste and reuse of produced waste. The model developed by the hospital is part of the organization's 'Quality Management Plan' constituted by five items: patient safety, health programs accreditation, integrated management system for workers, campaigns/projects and the Joint International Commission (accreditation organization on international quality criteria). The indicators developed for the model focus on waste generation and consumption of process inputs. Communication and monitoring through sustainability reports are regarded as crucial for its success.

The 'Global Green and Healthy Hospitals Agenda' represents another example of framework applied in Brazil. The model was created by the coalition '*Salud Sin Daño*' (Health Care without Harm), composed by an international network of healthcare systems, hospitals, communities, unions and environmental organizations. Its main goal is to transform the worldwide healthcare service without negatively affecting patient care. It seeks environmental sustainability so that the healthcare sector is no longer a source of losses for people and environment. The framework has ten main objectives related to leadership, chemical substances, waste, electricity, transport, food, pharmaceutical products and facilities (GGHHA

2012). According to numbers related to the initiative, the Brazilian hospitals use large amounts of electricity that represent more than 10 % of the total of the commercial energy consumption in the country. By comparison, the UK NHS carbon footprint represents more than 18 million tons of CO_2 per year, which is equivalent to 25 % of the total of public sector emissions (GGHHA 2012). The global agenda of the framework presents a comprehensive list of objectives and actions that can be implemented in the search for performance improvement, but does not present guidelines on how to find specific indicators. There are recommendations for the development and implementation of measures from documents that report experiences of hospitals, governmental entities and research in the field. These documents are available in their website and labelled as 'Tools and Resources'.

Current legislation has also been used to compose frameworks for the healthcare sustainability measurement. That is the case with the Manual of Hospital Accreditation (Ministry of Health (MS) 2002). The document was inspired by the Latin America and Caribbean Hospital Accreditation Manual (Novaes and Paganini 1992). The manuscript, however, represents a very incipient guide to present meaningful indicators and/or metrics adjusted to the healthcare environmental measurement. Three main aspects are controlled: general maintenance, waste and water potability. The existence of indicators is recommended without systematic guidance.

The framework provided by ISO 14001 (ISO 14031 for performance evaluation) constitutes another model used in the healthcare services in Brazil. The standard offers some principles that are periodically verified and evaluated for the re-accreditation. ISO 14001 aims to promote the environmental accreditation of organizations. The Environmental Failure Mode and Effects Analysis (EFMEA) methodology is regarded as a traditional option for the search for flaws in product and/or process projects within ISO (Zambrano and Martins 2007). Regarding the standard and environmental sustainability, it corresponds to the calculation of the risk of the occurrence of a flaw through the multiplication of severity, occurrence and detection indices. Aspects such as the pollution of waters, lands and air are considered. In an industrial environmental, for instance, these aspects are considered: the inputs and outputs of the productive process, the types of flaws that can happen and existing controls. At the end of the process, the more prevalent risks and actions are identified. Brazilian Hospital Albert Einstein (HIAE) with its five units located in São Paulo represents a certified organization. In the organization, the creation and use of an Environmental Management System has built commitment to environmental issues; a more rational use of water, electricity and resources has been noticed. There has been an acknowledgement of the organization and its employees' environmental responsibility (Sociedade Beneficente Israelita Brasileira (SBIB) 2012).

The Global Reporting Initiative (GRI) methodology is based on a voluntary initiative to standardize sustainability reports. Indicators related to the three dimensions of sustainability are suggested within its guidelines that are also adopted by HIAE. Regarding the environmental dimension, indicators are listed and goals are monitored in terms of (1) materials, (2) energy, (3) water, (4) biodiversity, (5) emissions (6) effluents and waste, (7) products and services, (8) compliance, (9) transport, (10) overall (expenditure and in investments in environmental initiatives), (11) supplier environment assessment and (12) environmental grievance mechanisms. As reported by Morhardt et al. (2002), the indicators proposed by GRI encompass essential aspects of environmental performance. Moreover, GRI guidelines have been promoted by the United Nations Environment Programme (UNEP). The framework is regarded as the most prominent reporting guide due to its focus that goes beyond the environment performance measurement to include a balanced approach of financial (economic) and social issues leading to sustainability.

Considering the current use of the five frameworks previously mentioned, that is, the Balanced Scorecard (BSC), Global Green and Healthy Hospitals Agenda (GGHHA), the standard established by ISO (ISO 14031) and the Hospital Accreditation Manual (HAM) for the measurement of environmental performance, a comparison between these models was made. Table 4 lists key aspects highlighted in literature as crucial for a meaningful measurement of environmental performance while looking for a correspondence in these frameworks.

As per the discussion on metrics of environmental sustainability performance and the analysis of current frameworks, a gap in research can be perceived. Performance metrics should be relevant and incorporate certain characteristics necessary to measure performance in healthcare organizations. The troublesome

Table 4 Key aspects of environmental performance measurement vs. current frameworks

Key aspects	GRI	ISO	GGHHA	HAM	BSC
Indicators based upon process measures (relevance, robustness and applicability)	2	2	1	1	1
Performance indicators associated with measurability, validity and controllability	2	1	1	1	1
Emphasis on the importance of multiple stakeholders	2	0	2	1	1
Definition of involved actors, target audience and level of performance to be measured	1	1	1	1	0
A sufficient number of performance indicators in order to capture the richness and scope of the organizational processes	2	2	1	1	1
Developed measures present meaning for the process being evaluated	2	2	1	2	1
An initial step/procedure is included to evaluate the current sustainability state	1	1	1	1	1
A sufficiently systematic and comprehensive methodology is presented, focusing on required characteristics for indicators and measures	0	0	0	0	0

Scale: 0—'not at all'; 1—'a little'; 2—'some'

deployment of measures and indicators within the strategic, tactical and operational levels is also underlined. Moreover, two perspectives should be present while developing measures: resource and value. Resources refer to the evaluation of the use of productive resources: increase and decrease. Value corresponds to the functional performance of each resource within its measure. Furthermore, the way performance data is validated has been discussed. It should involve stakeholders in the development of metrics to improve understanding and emphasize the importance of the measurement. Multiple points of view should be considered to define a sufficient number of performance indicators and metrics. Finally, the robustness of indicators/metrics is a vital milestone for the performance measurement process. Such robustness encompasses three main components: reliability, validity and adjustability. Section 4 describes the main findings of the bibliometric review of literature.

4 The Bibliometric Review

4.1 Search and Analysis Strategy

As previously stated, four databases were selected for the search: Web of Science, Medline Complete (EBSCO), Wiley Online Library and Scopus (Elsevier). The following terms were combined and sought in the Abstracts: 'Hospital', 'Operations', 'Environment*', 'Indicators', 'Performance', 'Sustain*', 'Framework', 'Quality', 'Health' and 'Care'. These terms were used due to their high occurrence identified in the initial literature review. The study analyzed published work between 1988 and 2013, i.e., 25 years since the publication of the Brundtland Report (Barkemeyer et al. 2014). Figure 1 represents the search and analysis strategy.

The combination of the terms: 'Health', 'Care', 'Performance', 'Indicators', 'Environment', 'Sustain*' produced 49 articles that given their comprehensiveness and focus were regarded as suitable for a more in-depth content analysis. After removing duplicates, 35 papers[2] were selected.

[2] It is important to indicate that the set of 35 papers is listed at the end of the list of references and arranged alphabetically (Abou-Ali and Abdelfattah 2013; Aitken et al. 2011; Andrade et al. 2012; Boriani et al. 2013; Boyer et al. 2012; Braithwaite et al. 2012; Brown et al. 2013; Chen et al. 2013; El-Jardali et al. 2011; Freitas et al. 2011; Gajewski and Dunton 2013; Gimelli and Muccillo 2013; Hao et al. 2012; Harmancioglu et al. 2013; Iacone et al. 2012; Joiner and Coleman 2012; Klazinga et al. 2011; Koutkias et al. 2012; Lin et al. 2013; Mears et al. 2011; Menikpura et al. 2012; Mutale et al. 2013; Mutemwa 2005; Olafsdottir et al. 2011; Peterson et al. 2013; Santana-Medina et al. 2013; Seke et al. 2013; Sulku 2012; Tanaka et al. 2013; Tchouaket et al. 2012; Townend 1996; Tseng 2013; Wakai et al. 2013; Yi and Ma 2010; Yildiz and Demirors 2013). Within the text, the format author(s) name/year of publication was adopted for easiness.

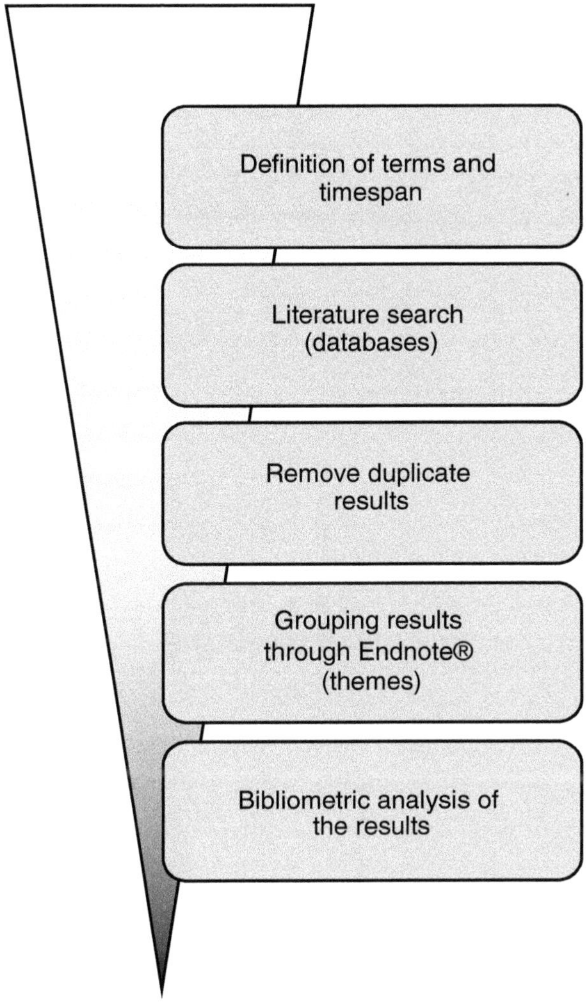

Fig. 1 The bibliometric review—search and analysis strategy

It is important to underline studies that appeared in more than one combination of terms: (Chen et al. 2013; Joiner and Coleman 2012; Klazinga et al. 2011; Wakai et al. 2013) and Gajewski and Dunton (2013) were found twice. Lin et al. (2013), Tanaka et al. (2013) were found three and four times, respectively.

By revealing impact of research, citation numbers have been considered as important indications of quality (Bornmann et al. 2008). Four main aspects were analyzed: (a) country of origin; (b) citation numbers; (c) publication outlets (journals, conferences); (d) main themes associated with environmental performance measurement of healthcare organizations. Table 5 shows the number of results obtained for each combination of terms (10 in total).

Table 5 The bibliometric review –results of the combination of terms

	Combination of terms	Web of Science	Medline Complete (Ebsco)	Wiley Online Library	Scopus (Elsevier)	Total
1	Health + Care + Performance	16,595	17,812	1,408	49,849	85,664
2	Hospital + Environment*	13,304	19,099	2,076	32,099	66,578
3	Hospital + Performance	20,873	2,088	2,318	39,495	64,774
4	Quality + Indicators + Health + Care	4,619	4,217	438	20,525	29,799
5	Indicators + Environment* + Sustain*	1,730	1,024	179	2,487	5,420
6	Health + Care + Operations + Performance	516	336	34	1,846	2,732
7	Hospital + Environment* + Indicator	449	313	70	1,090	1,922
8	Hospital + Indicator + Framework	203	213	25	422	863
9	Hospital + Sustain* + Environment* + Framework	24	20	8	49	101
10	Health + Care + Performance + Indicators + Environment* + Sustain*	12	7	7	23	49

Regarding country of origin, a certain prevalence of collaborative studies was noticed. The UK, the US and Turkey are present in more than two occasions. Nonetheless, the review of the papers does not present sufficient data for an account of prevalence of a country or research center. Table 6 displays the information on country of origin.

Given the reduced number of citations found in the first search within Web of Science, a new data collection effort was made through Google Scholar®. This search mechanism is regarded as comprehensive by including peer-reviewed papers, theses, books, abstracts and academic papers published by professional organizations, pre-publication libraries, universities and other academic entities. Figure 2 describes the results of both searches.

After comparing the results, a significant increase was perceived. Nevertheless, the most cited papers remained the same: Mutemwa, Klazinga et al., Aitken et al. (2011), Andrade et al. (2012). Mumtemwa investigates the effectiveness of Information Systems for healthcare management in Zambia. The study was associated with the Centre for AIDS Research of the University of Southampton in the United Kingdom. It does not, however, address environment sustainability; it focuses on the economic and social impact of technology to forge sustainability.

The paper by Klazinga et al. listed a number of themes related to the quality of healthcare results. The need for more valid and reliable performance indicators was pointed out in the research. Aitken et al. (2011) focused on human resources and the internal environment of hospitals in terms of their role to achieve sustainability. Andrade et al. (2012) explored the physical and social environment of four Portuguese hospitals: two old units and two recently renovated hospitals. The main purpose behind the study was to develop quality evaluation mechanisms for internal settings.

Table 6 The bibliometric review – countries of origin

Australia	*Japan*	*Turkey*
Aitken et al. (2011)	Tanaka et al. (2013)	Harmancioglu et al. (2013)
Australia, Canada, Ireland,	*Lebanon*	*Yildiz and Demirors* 2013*UK*
South Africa	Ei-Jardali et al. (2011)	Townend (1996)
Braithwaite et al. (2012)	*Mexico, France*	*UK, Malasia*
Canada	Santana-Medina et al. (2013)	Olafsdottir et al. (2011)
Tchouaket et al. (2012)	*Portugal*	*UK, The Netherlands*
China	Freitas et al. (2011)	Mears et al. (2011)
Denmark, Italy	*Portugal, Italy*	*US*
Boriani et al. (2013)	Andrade et al. (2012)	Boyer et al. (2012)
Egypt	*Serbia*	Joiner and Coleman (2012)
Germany, China, Japan	Seke et al. (2013)	Peterson et al. (2013)
Lin et al. (2013)	*Sweden, China*	*US, UK*
Greece, France, Denmark	*Tailand*	*Zambia, UK*
Koutkias et al. (2012)	Menikpura et al. (2012)	Mutale et al. (2013)
Ireland	*Taiwan*	
Wakai et al. (2013)	Chen et al. (2013)	
Italy	Tseng (2013)	
Gimelli and Muccillo (2013)	*The Netherlands*	

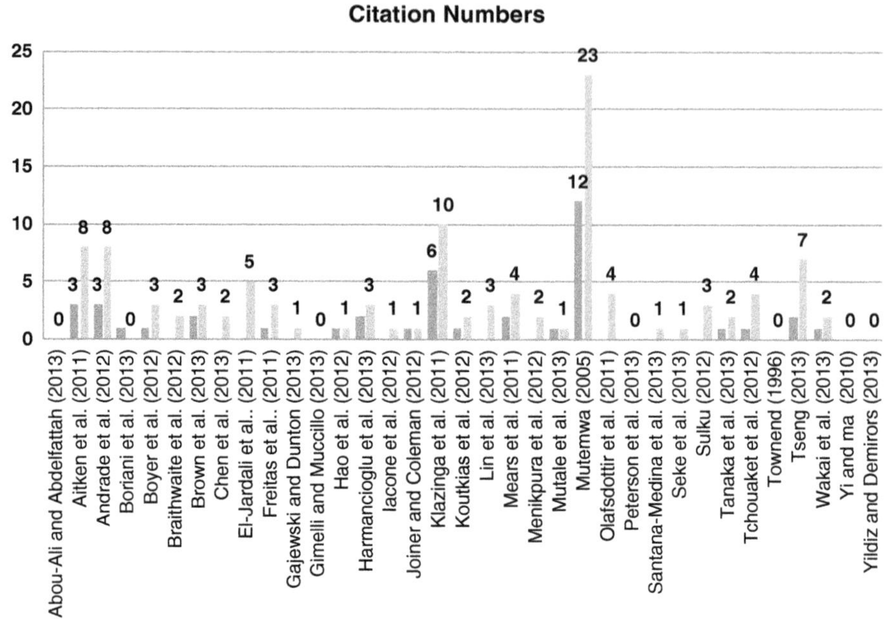

Fig. 2 The bibliometric review—citation numbers (Web of Science/Google Scholar)

The publication outlets were also analyzed. In only three cases, publications were cited more than once within the list. Two papers were published within BMC Public Health (Impact Factor: 2.32); two articles were published in the International Journal of Quality in Health Care (Impact Factor: 1.584) and two papers were published within Water Resources Management (Impact Factor: 2.463). The information on Impact Factors was collected from their websites. Two of the analyzed articles were published in conference proceedings: (Freitas et al. 2011; Tchouaket et al. 2012). Table 7 lists the papers by publication outlet.

As previously stated, the bibliometric review also aimed to analyze the main themes being discussed in the context of the healthcare organizations. Figure 3 represents the analysis on emerging themes. The coding of different themes followed the sequence: Paper title-Abstract-Keywords. Nine main themes were coded in the analysis. Three studies focused on the development of quality indicators based on benchmarking. Four papers emphasized the importance of 'best practices' in terms of quality improvement in the internal environment of hospitals. Six articles focused on strategies to improve economic/social efficiency of healthcare organizations (based on the Balanced Scorecard). One study focused on cleaner production in general. Economic development, especially at country and city level, was evaluated in three of the studies. Five papers reported on initiatives to improve efficiency in healthcare, mainly connected with cost reduction and economic savings of hospital units. One paper discussed Hospital Accreditation. Two of the studies were related to the role/impact of Information Systems to improve

Table 7 The bibliometric review—publication outlets (main journals)

7th ISWA International Congress, Proceedings II	**Informatics for Health & Social Care** Yildiz and Demirors (2013)
Townend (1996)	
Academic Medicine	
Joiner and Coleman (2012)	Freitas et al. (2011)
Advances in Management of Technology, Pt 1	**International Journal for Quality in Health Care**
	Braithwaite et al. (2012)
Applied Economics	Mears et al. (2011)
	International Journal of Health Planning and Management
Applied Energy	Tchouaket et al. (2012)
Gimelli and Muccillo (2013)	**Journal of Biomedical Informatics**
Australian Critical Care	Koutkias et al. (2012)
Aitken et al. (2011)	**Journal of Cleaner Production**
BMC Public Health	Tseng (2013)
Olafsdottir et al. (2011)	**Journal of Environmental Psychology**
Seke et al. (2013)	Andrade et al. (2012)
Building and Environment	**Journal of Evaluation in Clinical Practice**
	Tanaka et al. (2013)
Ecological Indicators	**Journal of Health Services Research & Policy**
Santana-Medina et al. (2013)	
Economic Modelling	**Journal of Operations Management**
	Boyer et al. (2012)
European Journal of Emergency Medicine	**Medical Decision Making**
Wakai et al. (2013)	
European Journal of Public Health	**Plos One**
	Mutale et al. (2013)
Expert Systems with Applications	**Quality & Quantity**
Lin et al. (2013)	Chen et al. (2013)
Frontiers of Environmental Science and Engineering	**Science of the Total Environment** Boriani et al. (2013)
Hao et al. (2012)	**Waste Management & Research**
Health Policy	Menikpura et al. (2012)
Ei-Jardali et al. (2011)	**Water Resources Management**
Health Policy and Planning	Harmancioglu et al. (2013)
	Peterson et al. (2013)

healthcare performance and quality. And ten of the articles reported on indicators of the preservation of natural resources, biodiversity and reduction of solid waste.

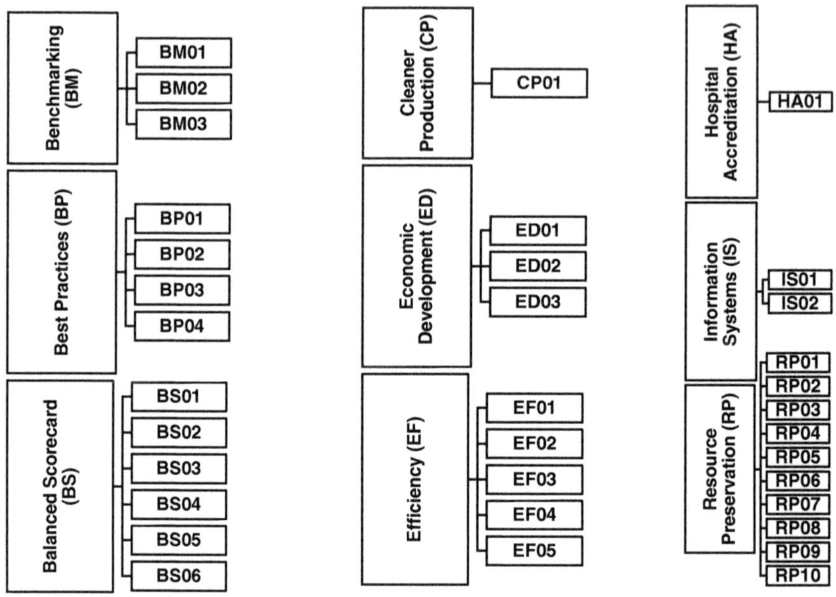

Code	References	Code	References	Code	References
BM01	Freitas *et al.* (2011)	CP01	Tseng (2013)	HA01	Braithwaite et al. (2012)
BM02	Joiner and Coleman (2012)	ED01	Seke *et al.* (2013)	IS01	Mutemwa (2005)
BM03	Klazinga et al. (2011)	ED02	Tchouaket et al. (2012)	IS02	Yildiz and Demirors (2013)
BP01	Aitken *et al.* (2011)	ED03	Yi and ma (2010)	RP01	Abou-Ali and Abdelfattah (2013)
BP02	Andrade *et al.* (2012)	EF01	Gajewski and Dunton (2013)	RP02	Boriani *et al.* (2013)
BP03	Boyer et al. (2012)	EF02	Iacone *et al.* (2012)	RP03	Brown *et al.* (2013)
BP04	Koutkias *et al.* (2012)	EF03	Lin *et al.* (2013)	RP04	Gimelli e Muccillo (2013)
BS01	Chen et al. (2013)	EF04	Tanaka *et al.* (2013)	RP05	Hao et al. (2012)
BS02	El-Jardali *et al.* (2011)	EF05	Wakai *et al.* (2013)	RP06	Harmancioglu et al. (2013)
BS03	Mears *et al.* (2011)			RP07	Menikpura et al. (2012)
BS04	Mutale *et al.* (2013)			RP08	Peterson *et al.* (2013)
BS05	Olafsdottir *et al.* (2011)			RP09	Santana-Medina *et al.* (2013)
BS06	Sulku (2012)			RP10	Townend (1996)

Fig. 3 The bibliometric review—main themes

5 Final Considerations and Further Research

The literature has shown that there is scarce evidence on the formalization of performance measures across organizations. In many cases, the systems already in place are ignored when new metrics are established. Regarding performance measurement frameworks, the lack of guidelines on the development and selection of performance indicators was also identified in the study. A consistent concern on the external public was indicated while internally the sustainability indicators are often composed by dimensions that are not comparable, restrict, and even insignificant. As a result, the application of frameworks for the environmental sustainability measurement has been far from satisfactory. Considering the importance of performance measurement for the healthcare sector, especially in developing countries, and the significant impact of its operations, there is a clear need for more robust/systematic frameworks guided by practice. New models are necessary to define relevant indicators for monitoring and performance evaluation in the sector. The search for measures that can raise awareness, continuous improvement and innovation in organizational processes has just begun. The study represents a starting point in proposing a meaningful framework to measure environmental sustainability in hospitals within the Brazilian setting.

Acknowledgment The researchers would like to acknowledge the support of 'UNIDAVI' through a PhD scholarship provided to Mrs. Blass.

References

Atkinson G (2000) Measuring corporate sustainability. J Environ Plann Manage 43(2):235–252

Barkemeyer R, Holt D, Preuss Lutz, Tsang S (2014) What happened to the development in sustainable development? Business guidelines two decades after brundtland. Sustain Dev 22 (1):15–32 (First published 2011)

Beischel ME, Smith KR (1991) Linking the shop floor to the top floor. Manag Account 73(4):25

Becker B (1997) Sustainability assessment: a review of values. Concepts and Methodological Approaches, World Bank, Washington, DC

Bertrand W, De Vries G (2005) Lessons to be learned from operations management. Health Operations Management: patient flow logistics in health care, London, Routledge

Bossel H (1999) Indicators for Sustainable development: theory, method, applications: a report to the balaton group. IISD, Winnipeg

Bio2 Sustentabilidade (2012) Projeto de Gestão Ambiental—Hospital Sírio Libanês. Available at http://www.agendasustentavel.com.br/Case.aspx?id=555. Accessed 1 March 2012

Bornmann L, Mutz R, Neuhaus C, Daniel HD (2008) Citation counts for research evaluation: standards of good practice for analyzing bibliometric data and presenting and interpreting results. Ethics Sci Environ Politics 8(1):93–102

Bourne M, Mills J, Wilcox M, Neely A, Platts K (2000) Designing, implementing and updating performance measurement systems. Int J Oper Prod Manag 20(7):754–771

Brown MG (1996) Keeping score: using the right metrics to drive world-class performance. Quality Recourses, New York

Castañeda-Méndez K, Mangan K, Lavery AM (1998) The Role and application of the balanced scorecard in healthcare quality management. J Healthc Qual 20(1):10–13

Campbell B, Sayer JA, Frost P, Vermeulen S, Ruiz-Pérez M, Cunningham A, Prabhu R (2001) Assessing the performance of natural resource systems. Conservation Ecology 5 (2), 22, [online]. Available at http://www.consecol.org/vol5/iss2/art22/. Accessed 1 April 2012

Campos LMS, Selig PM (2002) SGADA—Sistema de gestão e avaliação do desempenho ambiental: a aplicação de um modelo de SGA que utilize o Balanced Scorecard (BSC). Revista Eletrônica de Administração (REAd) 8(6):1–23

Cross K, Lynch R (1989) Accounting for competitive performance. J Cost Manag 3(1):20–28

Curtright JW, Stolp-Smith S, Edell ES (2000) Strategic performance management: development of a performance measurement system at the Mayo Clinic. J Healthc Manag 45(1):58–68

Coote A (2002) Claiming the Health Dividend. King's Fund, London

Da Silva CE, Hoppe AE, Ravanello MM, Mello N (2005) Medical wastes management in the south of Brazil. Waste Manag 25(6):600–605

Department of Environment (DoE) (1990) The Environmental Protection Act. HMSO London

Dixon JR, Nanni AJ, Vollmann TE (1990) The new performance challenge: measuring operations for world-class competition. Dow Jones-Irwin, Homewood, Illinois

EFQM (2013) European Foundation for Quality Management. In: Fundamental Concepts. Available at http://www.efqm.org/efqm-model/fundamental-concepts. Accessed 10 Sep 2013

Environmental Protection Agency (2007) The lean and environment toolkit. Environmental Protection Agency, USA. Available at http://www.epa.gov/lean/toolkit/LeanEnviroToolkit.pdf. Accessed 1 July 2012

Feng SC, Joung CB (2011) A measurement infrastructure for sustainable manufacturing. Int J Sustain Manufact 2(2):204–211

Fiksel J, McDaniel J, Mendenhall C (1999) Measuring progress towards sustainability principles, process and best practices. Battelle Memorial Institute, Ohio

Fitzgerald L, Brignall S, Silvestro R, Voss C (1991) Performance measurement in service businesses. Chartered Institute of Management Accountants, London

Forrest CB, Villagra VV, Pope JE (2006) Managing the metric vs managing the patient: the physician's view of pay for performance. The Am J Managed Care 12(2):83–85

GGHHA (2012) Global green and healthy hospitals agenda. Available at http://hospitalesporlasaludambiental.net/. Accessed 1 July 2012

Ghalayini AM, Noble JS, Crowe TJ (1997) An integrated dynamic performance measurement system for improving manufacturing competitiveness. Int J Prod Econ 48(3):207–225

Gomes CF, Yasin MM, Lisboa JV (2004) A literature review of manufacturing performance measures and measurement in an organizational context: a framework and direction for future research. J Manufact Technol Manag 15(6):511–530

Gunasekaran A, Patel C, Tirtiroglu E (2001) Performance measures and metrics in a supply chain environment. Int J Oper Prod Manag 21(1/2):71–87

Gunasekaran A, Spalanzani A (2012) Sustainability of manufacturing and services: investigations for research and applications. Int J Prod Econ 140(1):35–47

Gurd B, Gao T (2008) Lives in the balance: an analysis of the balanced scorecard (BSC) in healthcare organizations. Int J Prod Perform Manag 57(1):6–21

Hardi P, Pinter L (1995) Models and methods of measuring sustainable development performance. International Institute for Sustainable Development (IISD), Winnipeg, 301 p

Jarousse LA (2012) Environmental sustainability programs for hospitals. Hosp Health Netw 86 (1):33–40

Kanji GP, Sá PM (2003) Sustaining healthcare excellence through performance measurement. Total Qual Manag Bus Excellence 14(3):269–289

Kaplan RS, Norton DP (1992) The balanced scorecard—measures that drive performance. Harvard Bus Rev 70(1):71–79

Kaplan RS, Norton DP (2004) The strategy map: guide to aligning intangible assets. Strategy Leadersh 32(5):10–17

Keegan DP, Eiler RG, Jones CR (1989) Are your performance measures obsolete? Manag Account 70(12):45–50

Kleindorfer PR, Singhal K, Van Wassenhove LN (2005) Sustainable Operations Management. Prod Oper Manag 14(4):482–492

Krauthammer M, Nenadic G (2004) Term identification in the biomedical literature. J Biomed Inform 37(6):512–526

La Forgia GM, Couttolenc BF (2008) Hospital Performance in Brazil: the search for excellence. World Bank Publications, Washington DC

Linton JD, Klassen R, Jayaraman V (2007) Sustainable supply chains: an introduction. J Oper Manag 25(6):1075–1082

Maki J, Qualls M, White B, Kleefield S, Crone R (2008) Health impact assessment and short-term medical missions: a methods study to evaluate quality of care. BMC Health Service Research 8:121–128

Mannion R, Davies HTO (2002) Reporting health care performance: learning from the past, prospects for the future. J Eval Clin Pract 8(2):215–228

Maskell B (1989) Performance measures of world class manufacturing. Manag Account 67(5):32–33

McGlynn EA (1997) Six challenges in measuring the quality of health care. Health Aff 16(3):7–21

Medori D, Steeple D (2000) A framework for auditing and enhancing performance measurement systems. Int J Oper Prod Manag 20(5):520–533

Melnyk SA, Stewart DM, Swink M (2004) Metrics and performance measurement in operations management: dealing with the metrics maze. J Oper Manag 22(3):209–218

Ministry of Health (MS) (2002) Manual Brasileiro de Acreditação Hospitalar. Brazilian Manual of Hospital Accreditation, Series A, Standards and Technical Manuals 117, 3rd Edition, Reviewed and Updated, 109p

Morhardt JE, Baird S, Freeman K (2002) Scoring corporate environmental and sustainability reports using GRI 2000, ISO 14031 and other criteria. Corp Soc Responsib Environ Manag 9 (4):215–233

Nazar MW, Pordeus IA, Werneck MAF (2005) Gerenciamento de resíduos sólidos de odontologia em postos de saúde da rede municipal de Belo Horizonte, Brasil. Revista Panamericana Salud Publica 17(4):237–242

Neely A (ed) (2004) Business performance measurement: theory and practice. Cambridge University Press, Cambridge, 2002 (2004 edition)

Neely AD, Adams C, Kennerley M (2002) The performance prism: the scorecard for measuring and managing business success. Prentice Hall Financial Times, London

Neely A, Richards H, Mills J, Platts K, Bourne M (1997) Designing performance measures: a structured approach. Int J Oper Prod Manag 17(11):1131–1152

Nikolaou IE, Tsalis TA (2013) Development of a sustainable balanced scorecard framework. Ecol Ind 34:76–86

Novaes HM, Paganini JM (1992) Garantia de qualidade: acreditação de hospitais para a América Latina e o Caribe. Pan American Health Organization, Series Silos nº, Washington, DC 13

Oliveira JM (2003) A questão dos resíduos de serviços de saúde na administração hospitalar. EnANPAD 27, Atibaia/SP, Proceedings

Otley D (1999) Performance management: a framework for management control systems research. Manag Account Res 10(4):363–382

Porter ME (2010) What is value in health care? The N Engl J Med 363(26):2477–2481

Pinheiro De Lima E, Gouvea Da Costa SE, Angelis JJ, Munik J (2013) Performance measurement systems: a consensual analysis of their roles. Int J Prod Econ 146(2):524–542

Phillips PS, Holley K, Bates MP, Freestone NP (2002) Corby waste not: an appraisal of the UK's largest holistic waste minimisation project. Resour Conserv Recycl 36(1):1–31

Paju M, Heilala J, Hentula M, Heikkila A, Johansson B, Leong S, Lyons K (2010) Framework and indicators for a sustainable manufacturing mapping methodology. In: Simulation Conference (WSC), Proceedings of the 2010 IEEE Winter: 3411-3422

Scerri A (2010) Accounting for sustainability: implementing a residential emissions reduction strategy using an approach that combines qualitative and quantitative 'indicators' of sustainability. Manag Environ Qual Int J 21(1):122–135

Swayne L, Duncan WJ, Ginter PM (2006) Strategic management of health care organizations, 5th edn. Blackwell Publishing, Pastow 869p

Silva SRA, Gonçalves MA, Siqueira PC, Silveira CAC (2008) As decisões de investimento na Fundação Hospitalar de Minas Gerais e seus reflexos nos indicadores de qualidade. RAHIS—Revista de Administração Hospitalar e Inovação em Saúde, Belo Horizonte, 1(1):46–57

Sociedade Beneficente Israelita Brasileira (SBIB) (2012) Programa Einstein de Sustentabilidade, Einstein Sustainability Program. Available at http://www.einstein.br/sobre-a-sociedade/sustentabilidade/Paginas/programa-de-sustentabilidade.aspx. Accessed 3 June 2012

Tawfik-Shukor AR, Klazinga NS, Arah OA (2007) Comparing health system performance assessment and management approaches in the Netherlands and Ontario, Canada. BMC Health Serv Res 7:25–37

Townend WK, Cheeseman C, Edgar J, Tudor T (2009) Factors driving the development of healthcare waste management in the United Kingdom over the past 60 years. Waste Manage Res 27:362–373

Tudor TL (2007) Towards the development of a standardized measurement unit for healthcare waste generation. Resour Conserv Recycl 50:319–333

Tudor TL, Noonan CL, Jenkin LET (2005) Healthcare waste management: a case study from the National health service in Cornwall. Waste Manag 25(6):606–615

Tudor TL, Barr SW, Gilg AW (2008) A novel conceptual framework for examining environmental behavior in large organizations: a case study of the Cornwall National Health Service (NHS) in the United Kingdom. Environ Behav 40(3):426–450

Van Raan AFJ (1996) Advanced bibliometric methods as quantitative core of peer review based evaluation and foresight exercises. Scientometrics 36(3):397–420

Van Raan AFJ, Van Leeuwen TN (2002) Assessment of the scientific basis of interdisciplinary, applied research: application of bibliometric methods in nutrition and food research. Res Policy 31(4):611–632

Ventura KS, Reis LFR, Takayanagui AMM (2010) Avaliação do gerenciamento de resíduos de serviços de saúde por meio de indicadores de desempenho. Engenharia Sanitária e Ambiental 15(2):167–176

Veleva V, Hart M, Greiner T, Crumbley C (2003) Indicators for measuring environmental sustainability: a case study of the pharmaceutical industry. Benchmarking: an Int J 10(2):107–119

Vieira DS, Rodrigues SS (2013) Picoli RL (2013) Gestão Ambiental e Resíduos Hospitalares: Uma análise sobre a legislação vigente e a viabilidade de implantação dos Econcentros e ECTE no Distrito Federal. IV Congresso Brasileiro de Gestão Ambiental, Salvador/BA, pp 25–28

Vilaça WPT, Oliveira MM (2008) Sustentabilidade e Comunicação no contexto hospitalar: estabelecendo a necessária conscientização. IX Congreso Latinoamericano de Investigación De La Comunicación, México, October

Webster J, Watson RT (2002) Analyzing the past to prepare for the future: writing a literature review. Manag Inf Syst Q 26(2):3

Wallin JA (2005) Bibliometric methods: pitfalls and possibilities. Basic Clin Pharmacol Toxicol 97(5):261–275

Weir E, D'Entremont N, Stalker S, Kurji K, Robinson V (2009) Applying the balanced scorecard to local public health performance measurement: deliberations and decisions. BMC Public Health 8(9):127–134

Wongrassamee S, Gardiner PD, Simmons JEL (2003) Performance measurement tools—the balanced scorecard and the EFQM excellence model. Measuring Bus Excellence 7(1):14–29

Woolridge A, Morrissey A, Phillips PS (2005) The development of strategic and tactical tools, using systems analysis for waste management in large, complex organisations: a case study in UK Health Care. Res Conserv Policy 44(2):115–137

Wilcock PM, Campion-Smith C (1998) Never mind the quality: feel the improvement. Qual Health Care: QHC 7(4):181

Zambrano TF, Martins MF (2007) Utilização do método FMEA para avaliação do risco ambiental. Gestão & Produção [online] 14(2):295–309

Zhu Q, Sarkis J (2004) Relationships between operational practices and performance among early adopters of green supply chain management practices in Chinese manufacturing enterprises. J Oper Manag 22(3):265–289

References—Bibliometric Review

Abou-Ali H, Abdelfattah YM (2013) Integrated paradigm for sustainable development: a panel data study. Econ Model 30:334–342. ISSN 0264–9993

Aitken LM, Hackwood B, Crouch S, Clayton S, West N, Carney D, Jack L (2011) Creating an environment to implement and sustain evidence based practice: a developmental process. Aust Crit Care 24(4):244–254. ISSN 1036–7314

Andrade C, Lima ML, Fornara F, Bonaiuto M (2012) Users' views of hospital environmental quality: validation of the Perceived Hospital Environment Quality Indicators (PHEQIs). J Environ Psychol 32(2):97–111. ISSN 0272–4944

Boriani E, Benfenati E, Baderna D, Thomsen M (2013) Application of ERICA index to evaluation of soil ecosystem health according to sustainability threshold for chemical impact. Sci Total Environ 443:134–142. ISSN 0048–9697

Boyer KK, Gardner JW, Schweikhart S (2012) Process quality improvement: an examination of general vs. outcome-specific climate and practices in hospitals. J Oper Manag 30(4):325–339. ISSN 0272–6963

Braithwaite J, Shaw CD, Moldovan M, Greenfield D, Hinchcliff R, Mumford V, Kristensen MB, Westbrook J, Nicklin W, Fortune T, Whittaker S (2012) Comparison of health service accreditation programs in low-and middle-income countries with those in higher income countries: a cross-sectional study. Int J Qual Health Care 24(6):568–577. ISSN 1353–4505

Brown NW, Malmqvist T, Bai W, Molinari M (2013) Sustainability assessment of renovation packages for increased energy efficiency for multi-family buildings in Sweden. Build Environ 61:140–148. ISSN 0360–1323

Chen HC, Chang TC, Lin YT, Cheng HY, Chang TL (2013) Application of DMAIC process to enhance health effects in caring institution. Qual Quant47(4):2065–2080. ISSN 0033–5177

El-Jardali F, Saleh S, Ataya N, Jamal D (2011) Design, implementation and scaling up of the balanced scorecard for hospitals in Lebanon: policy coherence and application lessons for low and middle income countries. Health Policy 103(2):305–314. ISSN 0168–8510

Freitas A, Costa T, Marques B, Gaspar J, Gomes J, Lopes F, Lema I (2011) A framework for the production and analysis of hospital quality indicators. In Information Technology in Bio-and Medical Informatics, Springer, Berlin, pp 96–105. ISSN 0302–9743

Gajewski BJ, Dunton N (2013) Identifying Individual Changes in Performance With Composite Quality Indicators While Accounting for Regression to the Mean. Medical Decision Making 33 (3):396–406. ISSN 0272-989X

Gimelli A, Muccillo M (2013) Optimization criteria for cogeneration systems: multi-objective approach and application in an hospital facility. Appl Energy 104:910–923. ISSN 0306–2619

Hao T, Du PF, Gao Y (2012) Water environment security indicator system for urban water management. Frontiers Environ Sci Eng 6(5):678–691. ISSN 2095–2201

Harmancioglu NB, Barbaros F, Cetinkaya CP (2013) Sustainability issues in water management. Water Resour Manag 27(6):1867–1891. ISSN 0920–4741

Iacone F, Martin S, Siciliani L, Smith PC (2012) Modelling the dynamics of a public health care system: evidence from time-series data. Appl Econ 44(23):2955–2968. ISSN 0003–6846

Joiner KA, Coleman DL (2012) Perspective: key indicators in academic medicine: a suggested framework for analysis. Acad Med 87(2):230–235. ISSN 1040–2446

Klazinga N, Fischer C, Asbroek A (2011) Health services research related to performance indicators and benchmarking in Europe. J Health Serv Res Policy 16:38–47. ISSN 1355–8196

Koutkias V, Kilintzis V, Stalidis G, Lazou K, Niès J, Durand-Texte L, McNair P, Beuscart R, Maglaveras N (2012) Knowledge engineering for adverse drug event prevention: on the design and development of a uniform, contextualized and sustainable knowledge-based framework. J biomed inform 45(3):495–506. ISSN 1532–0464

Lin QL, Liu L, Liu HC, Wang DJ (2013) Integrating hierarchical balanced scorecard with fuzzy linguistic for evaluating operating room performance in hospitals. Expert Syst Appl, 40 (6):1917–1924. ISSN 0957–4174

Mears A, Vesseur J, Hamblin R, Long P, Den Ouden L (2011) Classifying indicators of quality: a collaboration between dutch and english regulators. Int J Qual Health Care 23(6):637–644. ISSN 1353–4505

Menikpura SNM, Gheewala SH, Bonnet S (2012) Framework for life cycle sustainability assessment of municipal solid waste management systems with an application to a case study in Thailand. Waste Manag Res 30(7):708–719. ISSN 0734–242X

Mutale W, Godfrey-Fausset P, Mwanamwenge MT, Kasese N, Chintu N, Balabanova D, Spicer N, Ayles H (2013) Measuring health system strengthening: application of the balanced scorecard approach to rank the baseline performance of three rural districts in Zambia. PloS one 8(3): e58650. ISSN 1932–6203

Mutemwa RI (2005) HMIS and decision-making in Zambia: re-thinking information solutions for district health management in decentralized health systems. Health Policy Plann 21(1):40–52. ISSN 0268–1080

Olafsdottir AE, Reidpath DD, Pokhrel S, Allotey P (2011) Health systems performance in sub-Saharan Africa: governance, outcome and equity. BMC public health 11(1):237. ISSN 1471–2458

Peterson HM, Nieber JL, Kanivetsky R, Shmagin B (2013) Water resources sustainability indicator: application of the watershed characteristics approach. Water resour manag 27 (5):1221–1234. ISSN 0920–4741

Santana-Medina N, Franco-Maass S, Sánchez-Vera E, Imbernon J, Nava-Bernal G (2013) Participatory generation of sustainability indicators in a natural protected area of Mexico. Ecol Ind 25:1–9. ISSN 1470–160X

Seke K, Petrovic N, Jeremic V, Vukmirovic J, Kilibarda B, Martic M (2013) Sustainable development and public health: rating European countries. BMC public health 13(1):77. ISSN 1471–2458

Sulku SN (2012) The health sector reforms and the efficiency of public hospitals in Turkey: provincial markets. Eur J Public Health 22(5):634–638. ISSN 1101–1262

Tanaka M, Lee J, Ikai H, Imanaka Y (2013) Development of efficiency indicators of operating room management for multi-institutional comparisons. J eval clin prac 19(2):335–341. ISSN 1356–1294

Tchouaket ÉN, Lamarche PA, Goulet L, Contandriopoulos AP (2012) Health care system performance of 27 OECD countries. The Int J Health Plann Manag, 27(2):104–129. ISSN 0749–6753

Townend B (1996) The implications of sustainable development and the zero option for health care waste. 7th Iswa International Congress, Proceedings II:II74–II81

Tseng ML (2013) Modeling sustainable production indicators with linguistic preferences. J Cleaner Prod 40:46–56. ISSN 0959–6526

Wakai A, O'Sullivan R, Staunton P, Walsh C, Hickey F, Plunkett PK (2013) Development of key performance indicators for emergency departments in Ireland using an electronic modified-Delphi consensus approach. Eur J Emerg Med 20(2):109–114. ISSN 0969–9546

Yi R, Ma ZX (2010) Research on the comprehensive evaluation of small town sustainable development based on DEA. Adv Manag Technol Pt 1:264–270. ISBN 978-1-921712-08-1

Yildiz O, Demirors O (2013) Measuring healthcare process quality: applications in public hospitals in Turkey. Inform Health Soc Care 38(2):132–149. ISSN 1753–8157

How Does Innovativeness Foster Sustainable Supply Chain Management?

Jury Gualandris and Matteo Kalchschmidt

Abstract Innovativeness refers to an organization's willingness/ability to change through architectural/radical innovation of processes, products and management systems. Existing literature generally supports the importance of innovativeness for environmental and social sustainability. Open questions, however, still remain: (i) which mechanisms (i.e., tools and routines) do innovative firms leverage that better assist the process through which stakeholder pressure is turned into sustainability strategy? (ii) Which mechanisms do innovative firms leverage that better assist the process through which strategy is turned into a successful adoption of sustainable supply chain management (SSCM) practices? To advance our understanding of how innovativeness fosters SSCM, we leveraged a continuous process of theory generation and data collection through case studies. A set of propositions and a conceptual model were built that complement prior research and can provide guidance to firms that struggle with how to deliver new environmental and social standards in their supply chains.

Keywords Innovation · Sustainability · Stakeholder pressure · Ethical responsibility · Inertia · Trade-offs

1 Introduction

Sustainable supply chain management (SSCM) has been advocated as a new innovation for firms to meet stakeholders' requirements while raising ecological efficiency and social responsibility in supply chains (Ahi and Searcy 2013;

J. Gualandris (✉)
Michael Smurfit Graduate Business School, University College Dublin,
Carysfort Ave, Blackrock, Dublin, Ireland
e-mail: jury.gualandris@ucd.ie

M. Kalchschmidt
Department of Management, Information and Production Engineering,
Università degli Studi di Bergamo, Viale Marconi 5, Dalmine, Bergamo, Italy
e-mail: matteo.kalchschmidt@unibg.it

A. Chiarini (ed.), *Sustainable Operations Management*,
Measuring Operations Performance, DOI 10.1007/978-3-319-14002-5_6

Gualandris and Kalchschmidt 2014; Zhu et al. 2005). SSCM tackles environmental and social issues using supply chain management practices and can be viewed at multiple levels including internal practices (e.g., ISO 14001, OHSAS 18001) and external practices (e.g., supplier assessment and collaboration) (Darnall et al. 2008; Gualandris and Kalchschmidt 2014; Zhu et al. 2012). During the last two decades a growing stream of literature has been studying how SSCM develops (Seuring and Muller 2008). At a first stage, pressures from primary (i.e., customers) and secondary (i.e., governments, non-governmental organizations) stakeholders influence a firm's set of environmental and social objectives and plans (Ateş et al. 2011; Ehrgott et al. 2011; Sharma and Henriques 2005). At this stage, environmental and social values of individuals operating at different levels within the organization also play a significant role, influencing the formulation of a firm' objectives and plans (Carter and Jennings 2004; Ehrgott et al. 2011; Klassen 2001). At a second stage, once pressure is accommodated into strategy, effort concentrates on translating objectives and plans into successful adoption of SSCM practices (Handfield et al. 1997; Maignan et al. 2002; van Tulder et al. 2009).

In the course of the discussion on SSCM, it has also been widely acknowledged that innovation is of utmost importance along the abovementioned process. Several studies show that environmental and social harm produced by industrial systems must decrease 50 times (Isaksson et al. 2010) and be also complemented by positive or regenerative impacts (Pagell and Shevchenko 2014). Such leaps in performance demand radical changes in firm's strategies and operations and prove the crucial role of innovation for sustainability. Evidence is indeed emerging as regards to the enabling role of innovativeness in the context of SSCM (e.g., Gualandris and Kalchschmidt 2014; Klassen and Vereecke 2012; Pagell and Wu 2009). Since the goal of innovative firms is to create new businesses or generate additional profit from better products and processes, they are assumed to be best at radically changing the way they think about products, processes and technologies and "treat sustainability as innovation's new frontier" (Nidumolu et al. 2009).

Although existing literature generally supports the importance of innovativeness for sustainability, there are a number of issues with its role in the context of SSCM. First, we have a limited understanding of how innovativeness influences the process through which stakeholders' pressure is accommodated in a firm's sustainability strategy; specifically, which mechanisms (i.e., tools and routines) do innovative firms leverage that better assist the process through which pressure is accommodated into strategy? Second, it is unclear how maintaining a strong (weak) innovation capability fosters (hampers) the process through which a sustainability strategy cascades down into successful action; specifically, which mechanisms do innovative firms leverage that better assist the process through which strategy is turned into successful SSCM adoption?

In moving a step forward towards filling these gaps in the literature, we have developed an assessment framework to capture (a) the perceived intensity of sustainability-related pressure, (b) the advancement level of a firm's sustainability strategy, (c) the extent of successful adoption of SSCM practices, and (c) the degree of a firm's innovativeness. Leveraging such framework in a continuous process of theory generation and data collection through a series of case studies, we have built a set of propositions and a conceptual model describing how innovativeness fosters SSCM. Also, we have identified a set of mechanisms that assist the development of advanced sustainability strategies and foster the adoption of SSCM practices. We argue that our findings complements prior research and also provide guidance to firms that struggle with how to deliver new environmental and social standards in their supply chains.

2 Background

This section provides details about key terminology and conceptualization of the factors considered in this paper as implemented by existing literature. We begin by describing the two-stage process through which pressure for sustainability is turned into SSCM adoption; main barriers to this process are also summarized. Then, we discuss the literature that has investigated the relationship between innovativeness and sustainability.

2.1 From Stakeholders' Pressure to SSCM Adoption

The literature has isolated the causal relationship that characterizes the way SSCM develops within a firm. At a first stage, external pressure and ethical responsibility tend to be encoded in a firm's sustainability strategy, or in its set of environmental and social objectives and plans (e.g., Klassen 2001). The role of 'external pressure' in inducing/motivating environmental and social sustainability at the firm level has been widely recognized. As managers seem to not discern the different sustainability demands coming from different stakeholder groups (Murillo-Luna et al. 2008), customers and societal stakeholders (Governments and Non-Governmental Organizations) were shown to influence a firm's strategy in similar ways. Escalating penalties, fines and legal costs have punctuated the importance of complying with legislation (Bansal and Roth 2000). Non-governmental organizations can leverage withholding and indirect usage strategies to pressure for environmental and social improvements (Sharma and Henriques 2005). In a similar fashion, customers' requests and requirements have been highly regarded by managers so to maintain the necessary support from these key stakeholder group (Ehrgott et al. 2011; Klassen and Vachon 2003; Zhu et al. 2007). Ethically motivated firms, then, develop environmental and social objectives and plans because it is the "right thing to do" (Bansal and Roth 2000). 'Ethical responsibility', which refers to the genuine intention and willingness of management and employees in a firm to be engaged in sustainability management (Carroll 1999), is instrumental to prioritize sets of resources (e.g., financial or human capital) for the achievement of sustainability (Klassen 2001).

Pressure can be perceived and accommodated in different ways. When firms address environmental and social issues, indeed, they face different strategic options; Some firms decide to pursue a 'resistant adaptation' to new regulations and requirements, whereas others approach the subject more strategically, adopting a 'total environmental and social quality' strategy (TESQ) (Handfield et al. 1997; Lee and Rhee 2007; van Tulder et al. 2009). According to the literature, developing this last option mainly means (i) considering environmental and social objectives that go beyond compliance as a top priority for the business, (ii) dedicating important budgets to improve ecological efficiency, health and safety or avoid sustainability related risks, (iii) clearly assigning responsibility for environmental and social matters to one or various people of the firm who are specialized in the matter of interest.

At a second stage, 'sustainability strategies' must cascade down into 'SSCM adoption'. Intuitively, as a firm's strategy moves from 'resistant adaptation' to 'TESQ', the extent of adoption of SSCM practices is expected to increase (Ateş et al. 2011; Handfield et al. 1997). As said, SSCM tackles environmental and social issues using different sets of practices. Sustainable process management (SPM) refers to a firm's institutionalization of internal environmental and social management practices that aim to reduce a firm's direct environmental and social impact. In essence, SPM entails the joint adoption of environmental management systems (ISO 14001) (Daily and Huang 2001; Darnall et al. 2008), environmentally friendly eco design (e.g., Design for Environment, Life cycle assessment) (Zhu and Sarkis 2004), health and safety certifications (Robson et al. 2007) and other social practices (e.g., codes of conduct, corporate social activities) (Zairi and Peters 2002). Practices that include transactions with supply chain partners have been seen as external SSCM. Since a firm is 'as sustainable as their suppliers' (Krause et al. 2009), research on external SSCM practices has mainly focused on Sustainable Supply Management (SSM). It specifically refers to two complementary sets of activities implemented at the firm level to assess and improve the environmental and social performance of suppliers (Ageron et al. 2012; Awaysheh and Klassen 2010; Lee and Klassen 2009): supplier assessment and collaboration. While supplier assessment comprises activities conducted to assess (and control) suppliers' sustainability performance, supplier collaboration entails a direct involvement with suppliers to improve environmental and social impacts of their products and operations.

The two-stage process depicted above (pressure → strategy; strategy → practice) provides an initial understanding of firms' behavior. Recent investigations, however, suggest that several barriers impede firms to approach pressure strategically and also hamper the adoption of SSCM practices. For instance, despite pressure and scrutiny by societal stakeholders (Legislators, Greenpeace, Friends of the Earth), many firms struggle with how to define and pursue TESQ strategies (United Nation Global Compact and Business for Social Responsibility 2010). Research suggested that "static mind-sets", "rigid organizational structures" and "lacks of supportive culture" is what prevent firms, especially large organizations, to pursue radical sustainability innovations and to grasp the idea that environmental and social performance can be improved while avoiding rebound effects (Ashford et al. 2012; Herring and Sorrell

2009; Pagell and Gobeli 2009) and improving productivity (Pagell et al. 2013; Porter and Van der Linde 1995; Wu and Pagell 2011). Furthermore, as SSCM requires lots of effort as well as resource endowments that often cross a firm's boundaries, poor employees' commitment and a lack of supplier relationship management can hamper the adoption of SPM and SSM (Gualandris et al. 2014; Walker et al. 2008).

2.2 *Innovativeness and Its Role in SSCM*

'Innovativeness' can be measured at multiple levels: at the product/process level or at the firm level. At the product/process, in accordance with Garcia and Calantone (2002), innovativeness refers to the degree of novelty of an innovation or the ability of an innovation to create a paradigm shift in technology and/or market structure. In this sense, the authors classify the concept of innovation along a continuum from incremental innovation (adaptation or marginal improvement of existing products and processes) to radical innovation (product/process innovation capable of creating a paradigm shift in technology and/or market structure of an industry). In parallel, innovativeness at the firm level refers to an organization's willingness/ability to change through architectural/radical innovation of processes, products and management systems (Gualandris and Kalchschmidt 2014; Klassen and Vereecke 2012). In this work, we study innovativeness at the firm level.

Innovativeness is a dynamic capability (Eisenhardt and Martin 2000; Reuter et al. 2010) that develops over time because of a firm's 'future orientation' and 'learning orientation'. Future orientation refers to a firm's inclination to analyze long-term developments of markets and technologies, as well as their implications for strategy and competitiveness (Ruff 2006). Such an inclination requires the organization to open its foresight, which in turn induces fresh ideas and rich information for an effective management of change (Rohrbeck and Gemünden 2011). Learning orientation refers to extent a firm is committed to learn, create a shared-vision and enrich its organizational memory (Sinkula et al. 1997). While commitment to learning means valuing and promoting learning for the long-term benefit of the system, shared-vision and memory refer to a firm's routines of spreading relevant information and knowledge among different departments and units (Calantone et al. 2002; Hult and Ketchen 2003; Verona 1999).

Limited evidence is emerging that innovativeness is an important capability that assists firms when turning pressures into successful SSCM adoption. Christmann (2000), for instance, has found that innovativeness is a complementary asset that enables companies to overcome the trade-off between costs and respect for the environment. Christmann (2004) has found that firms that are characterized by a capacity for innovation, an ability to absorb new information thanks to an educated workforce, and a widespread involvement of employees in the implementation of an environmental management system, are in a much better position to adopt strategies of environmental self-regulation such as the ISO14001 standard. Recent research has also found that innovative firms are better at sustainability, as they are able to create and leverage on strong supply chain participation (Bos-Brouwers 2010; Pagell and Wu 2009). Gualandris and Kalchschmidt (2014) have provided preliminary evidence that innovativeness enhances a firm's ability to identify and respond to emerging customer's requests and requirements. Likewise, based on their case studies, Klassen and Vereecke (2012) have found that innovation is a valuable capability for addressing social issues in supply chain operations.

Besides SSCM literature, nearby fields have developed some understanding of the relationship between innovativeness and sustainability. For instance, innovation management research presents several cases of innovative firms that are best at embracing and combining different sustainable business models (Bocken et al. 2014), thus proactively accommodating stakeholders' sustainability expectations. The business ethics literature, suggest that the ability of creating "thinking chambers" where a variety of stakeholders can debate problems and needs for change allows firms to "detect supply chain innovation opportunities for sustainable development" (Isaksson et al. 2010). Finally, sustainable entrepreneurship literature refers to innovative firms as those capable of "creative destruction" (Schumpeter 1934), i.e., they destroy conventional production methods and products, and replace them with safer and greener operations and services in order to create, deliver and capture social, environmental and economic value (Schaltegger and Wagner 2011).

The relationship between innovation and sustainability is far from being a novel subject. However, the role innovativeness plays along the two-stage process through which sustainability-related pressure is turned into successful SSCM adoption is far from being completely understood. Indeed, it remains unclear how innovativeness fosters the development of TESQ strategies and the subsequent adoption of SPM and SSM. This study, thus, concentrates on the complex interaction between innovativeness, pressure, strategy and practice, discerning light on the specific tools and routines that make innovative firms best at sustainability.

3 Methodology and Case Results

In this study we used case-based methodology for two reasons. First, case studies are widely considered to be better means of theory building where the relationships among important constructs are poorly understood. Second, the case-based method enables to analyze the phenomenon in its natural setting, based on firsthand observation and examination of numerous factors and nuances; accordingly, recent literature asked for case-based investigation of the role of innovativeness in the context of SSCM (van Bommel 2011).

In accordance with Stuart et al. (2002), five methodological steps were performed. First, research questions were identified (see previous sections); then, an interview protocol was developed; next, field and archival data were gathered;

fourth, interviews were transcribed and analysis performed using an assessment framework and cross-case comparisons; and finally, results and considerations were disseminated to improve validity.

3.1 Sample Selection and Data Collection

This study investigates the role of innovativeness along the process through which stakeholders' pressures are turned into SSCM adoption. Therefore, case selection was accurately managed to obtain a sample of firms facing similar pressure but having different strategies and practices in place. We focused on Italian manufacturing firms since they have recently faced increasing local pressure and several critical environmental and social challenges in the area of hazardous material substitution, health and safety, as well as recovery and recycling. This focus has also allowed controlling for potentially confounding factors such as variation in economic conditions and national culture. Then, we applied the 'polar types' sampling approach (Eisenhardt and Graebner 2007); extreme cases in terms of sustainability strategies were identified by considering the attitude of transparency on firms' websites as well as firms' membership in Italian corporate-social-responsibility (CSR) networks such as Fondazione Sodalitas (www.sodalitas.it) and Impronta Etica (www.improntaetica.org). Variance in SSCM adoptions was purposefully created by selecting cases on the basis of firms' environmental and social certifications. Table 1 briefly summarizes firms' profiles.

The unit of analysis for these case studies was the process through which pressure turns into strategy and then into practice. Case studies were developed by means of interviews conducted on the basis of a semi-structured protocol (Yin 2009). Such a protocol was designed to explore (i) how external pressure and ethical responsibility affect/motivate a firm's sustainability strategy, (ii) the level of advancement of a firm's sustainability strategy (resistant adaptation-TESQ), (iii) the extent of successful adoption of SSCM practices, (iv) a firm's degree of innovativeness, and (v) innovativeness-related mechanisms (i.e., tools and routines) enabling a firm to integrate sustainability into its strategy and practices.

Because studying the role of innovativeness in the context of SSCM requires the inclusion of multiple perspectives, in each firm, at least two individuals were interviewed (Table 1). All interviews were conducted on site and recorded electronically. Most of them have lasted for more than 90 min. After each site visit, each interviewer edited field notes and checked for accuracy. Questions arising from interview notes have been answered by interviewees through follow-up emails and phone calls.

Data gathered from different interviews have been triangulated by means of overlapping interviews, archival documents when available and sustainability reports. The continuous process of theory generation and data collection was leveraged until external saturation was reached (Yin 2009); during the last three interviews no new information or new themes were emerging from data analysis. Thus we stopped at 10 cases.

Table 1 Cases

Cases										
Items	A	B	C	D	E	F	G	H	I	J
Foundation	1955	2000	1966	1870	1964	1975	1836	1988	1899	1956
Industry and products	ISIC 34 light-heavy duty cranes	ISIC 29 weaving systems and healed frames solutions	ISIC 34 cabs, driver units for track loaders	ISIC 29 cutting machine tools and energy solutions	ISIC 29 pneumatic equipment for the automation	ISIC 31 domestic appliances	ISIC 31 network connectivity systems and circuit breakers	ISIC 31 network connectivity systems and circuit breakers	ISIC 32 home automation, and medical systems	ISIC 29 air cooling and conditioning equipment
Employees	450	545	648	283	383	1000	700	615	472	220
Membership in CSR networks	No	No	No	No	No	Yes	Yes	Yes	Yes	No
Certifications	ISO 9001	ISO 9001	ISO 9001, ISO 14001	ISO 9001	ISO 9001; ISO 14001; OHSAS 18001	ISO 9001; ISO 14001; OHSAS 18001; SA 8000	ISO 9001; ISO 14001; OHSAS 18001	ISO 9001; ISO 14001; OHSAS 18001	ISO 9001; ISO 14001; OHSAS 18001	ISO 9001; ISO 14001
Archival documents	Sustainable development statement (2011)	Environment obligation statement (2010)	Supplier questionnaire (2010)	Corporate governance code (2011)	Code of Ethics (2011)	Sustainability reports (2001–2011)	Sustainability reports (2003–2011)	Sustainability reports (2011); sustainability KPI indexes (2010–2011)	Sustainability KPI report (2010); suppliers' code of conduct (2011)	Energy saving brochure (2010)
Interviewees	Purchasing manager; plant manager	Operations manager; senior buyer	Purchasing manager; R&D manager	Operations manager; senior buyer	Plant manager; senior buyer	Strategic sourcing manager; technical affair manager; R&D manager	Purchasing manager; HR director	Sustainability director; supply chain director.	Purchasing manager; plant manager	Purchasing manager; R&D manager

3.2 Assessment Framework and Data Analysis

Drawing on the synthesis of earlier research (e.g., Calantone et al. 2002; Handfield et al. 1997; Klassen 2001), an assessment framework was developed for the fieldwork. Tables 3, 4, 5 and 6 respectively provide information about our assessment framework.

In our framework, each construct comprised a number of lower-level dimensions. Each lower-level dimension was rated separately on a scale of 1–3 (i.e., low, medium, high). Then, we calculated the average of lower-level dimensions to assess the construct of interest. To illustrate, consider 'SSCM adoption' (Table 6): we have two middle-level dimensions (SPM and SSM) and a number of bottom-level dimensions. For SPM, bottom-level dimensions are (1) hazardous material management within the product, (2) use of Eco-design (e.g., Design for Environment and Life Cycle Assessment), (3) building an environmental management system (EMS), (4) reduction of energy, water use and emission, (5) improving health and safety (e.g., obtaining OHSAS 18001), (6) improving corporate responsibility through social campaigns (e.g., codes of conduct, corporate social activities). Scores across the bottom-level dimensions are averaged based on the assumption that each contributed significantly and more or less equally to SPM. Then, scores for SPM and SSM are averaged to obtain the 'overall' extent of successful SSCM adoption.

Based on data collected from overlapping interviews, archival documents and sustainability reports, the research team members independently assessed the role of external pressure and ethical responsibility in inducing/motivating a firm's sustainability strategy, the type of sustainability strategy, the extent of successful adoption of SSCM practices, and the degree of innovativeness. Each independent assessment was then discussed collectively to resolve discrepancies; the final consensus-based ratings are reported in Table 2.

Data analysis based on our assessment framework was necessary for two main reasons (Lee and Klassen 2009); first, we wanted to limit interpretation bias in the single-case analysis and facilitate the process through which different pieces of

Table 2 Single-case reduction and cross-case comparison

Cases	1. Pressure				2. Strategy	3. SSCM adoption			Innovativeness		
	Overall	SP	CP	ER	Overall	Overall	SPM	SSM	Overall	FO	LO
A	2.56	2.67	2.50	2.50	1.00	1.42	1.83	1.00	1.44	1.00	1.89
B	2.50	3.00	2.00	2.50	1.50	1.50	1.50	1.50	1.43	1.20	1.67
C	2.50	3.00	2.50	2.00	1.50	1.67	1.83	1.67	1.69	1.60	1.78
D	1.50	2.00	1.50	1.00	1.00	1.08	1.17	1.17	1.60	1.20	2.00
E	2.17	2.00	2.00	2.50	2.50	2.00	2.17	1.83	1.96	1.80	2.11
F	1.61	1.33	1.00	2.50	3.00	3.00	3.00	3.00	3.00	3.00	3.00
G	1.78	1.33	1.50	2.50	3.00	2.83	2.67	2.67	2.79	2.80	2.78
H	1.72	1.67	1.00	2.50	2.50	2.58	2.67	2.50	2.72	3.00	2.44
I	1.33	1.00	1.00	2.00	3.00	2.92	2.83	3.00	2.90	2.80	3.00
J	1.72	1.67	1.50	2.00	2.00	2.17	2.33	2.17	2.32	2.20	2.44

information coming from different sources (interviews, archives and reports) were reconnected. Second, having quantitative scores allowed the research team to look at the big picture without losing detail, thus improving the reliability of the cross-case analysis. Specifically, Table 2 was useful to assess the coherence of pressure, strategy and SSCM adoption. Every time misalignment was identified (e.g., firm E obtaining high score for sustainability strategy, but low score for SSCM adoption), a meaningful cross-case comparison was conducted to assess the role of innovativeness.

4 Development of Propositions

Based on the two-stage process presented in Sect. 2, we separated the findings into two areas: turning pressure into strategy; turning strategy into action.

4.1 Turning Pressure into Strategy

Our analysis informs about the coherence of pressure and sustainability strategy, and allows assessing the role of innovativeness. Despite the studied cases were exposed to similar external pressure from primary and societal stakeholders (i.e., same industrial context), our analysis has shown that firms perceive and cope with such pressure differently. In certain cases (A, B, C and D), external pressure was turning straight into specific environmental and social actions in an attempt to limit legal liabilities and maintain business with important customers. In such cases, explicit requests and requirements from primary and societal stakeholders were seen as an essential source of information, uncovering sustainability meanings and shedding light on salient environmental and social issues that had be addressed urgently. Firms A, B, C and D appeared to be myopic: their inability to engage with stakeholders and predict potential outcomes of alternative strategic options resulted in a 'resistant adaptation' strategy.

In other cases (i.e., F, G and I), request and requirements from primary and societal stakeholders were not perceived as a salient source of pressure, as firms were already addressing environmental and social issues. Specifically, we observed that inclusive approaches based on a two-way communication with external and internal stakeholders were leveraged to inform the firm's strategic thinking, thus anticipating most of stakeholders' demand. To illustrate, the technical affair manager at firm F highlighted that "interactive dialogs with customers provide ideas about what to look for in business… when sustainable alternatives in products or technologies are identified, those are compared with current practices… If a realignment [of our strategy] is needed, our effort in collecting new ideas from employees and partners and our effort in keeping on discussions provide arguments that allow for taking a conscious decision". Firms F, G and I were good at creating constructive interactions between customers, suppliers and local communities in a multi-partite stakeholder cooperation from which stem mutual loyalty as well as better understanding of reciprocal expectations. It was clear in the mind of the interviewed managers that virtually no company should innovate on its own; "the new innovation leaders will be those finding the best way to leverage its network of stakeholders" (Manager at firm F). A standard routine for such firms was, indeed, to bring people and ideas together through 'innovation mall' and 'innovation community'. Firm G and I were circulating a weekly newsletter; a simple tool through which the firm could post problems and stakeholders could post ideas and solutions (i.e., innovation mall). Firm F was leveraging a more advanced tool, a blog where stakeholders could open discussions and discourses as well as offer solutions to problems posted by the company (i.e., innovation community). Beyond mutual loyalty and reciprocal understanding, such mechanisms allowed to leverage stakeholders' skills and capabilities, initiate market and social changes and finally deliver environmental and social innovations.

As regards to ethical responsibility, cases existed where it could not be turned into TESQ strategies (Table 2). In such cases (i.e., cases A, B and C), the diffusion/proliferation of top management's and employees' genuine interest towards sustainability was hampered because occasions for discussion and sperimentation were missing. Conversely, we observed cases where organizations implemented mechanisms that allowed their constituents to manifest their values, thus influencing a firm's objectives and plans (i.e., cases F, G and I). To illustrate, within Firm I 'ideas competitions' took place frequently and employees were allowed to use 10 % of their paid working time to pursue their intents. Employees were also receiving extra money in the form of 'reimbursements' when small investments were required in order to develop or refine their ideas/solutions. 'Open meetings' were also organized on a frequent-basis to discuss successful and unsuccessful endevours, and share knowledge within the organization. When firm I realized that social innovations conducted by ethically motivated employees were positively impacting its economic bottom line, a TESQ strategy with sustainability-related objectives, roles and budgets was formally constituted.

It is important to highlight that 'innovation malls', 'innovation communities', 'innovation ideas competitions', 'reimbursements' and 'open meetings' were not purposefully created to inform a firm's sustainability strategy or enhance environmental and social performance; firms G, I and F were mainly concerned about maintaining/enhancing their innovativeness. However, because of primary and societal stakeholders' self-interest as well as employees' ethical values, such mechanisms were often leveraged to discuss and tackle environmental and social issues. Collectively, our observation leads to the following:

Proposition 1 (P1) As innovativeness increases, a TESQ strategy is more likely to be formulated and pursued.

P1a. *As innovativeness increases, external pressure is less likely to affect a firm's sustainability strategy since organizations employ mechanisms (i.e., innovation mall* *and community) that allow to anticipate/shape explicit demands from primary and societal stakeholders.*

P1b. *As innovativeness increases, ethical responsibility is more likely to affect a firm's sustainability strategy since organizations employ mechanisms (i.e., ideas competitions, reimbursements, open meetings) that allow individuals to manifest, share and pursue their values.*

4.2 Turning Strategy into Action

We found a case where the TESQ strategy could not turn into extensive and successful adoption of SSCM practices: firm E had defined specific sustainability priorities, roles and budgets, but SSCM could not be implemented extensively and successfully (Table 2). Firm E had developed complex internal and external structures over time (e.g., mechanistic organization, regionally differentiated product portfolio, globally spread supply base), as a consequence of its internationalization process and increased size. We observed such structures to be source of inertia and friction, which impeded taken-for-granted beliefs and ways of operate to be disrupted for accommodating new operational paradigms (SSCM). For instance, the firm wanted to adopt new environmental systems and safety measures along its production lines; such new practices, however, were requiring significant effort to adjust workers' routines and skills, thus creating stress and negatively impacting their productivity. Yet, firm E was making an effort to involve Chinese suppliers along recycling and eco-design initiatives; cultural differences and communication barriers, however, were impeding the achievement of significant eco-efficiency gains. After few years, such responsible production and sourcing projects were still not completed.

Firm I and F had developed similar structures over time; for instance, the plant manager of firm F told us that "functional managers [in our organization] are subjected to pressures from different stakeholders and might develop potentially misaligned goals and practices [...] in the past, this was a big source of inefficiency, especially when the market required quick turn-around and adjustments in products and operations". However, as winning such inertia was essential to maintain market leadership, the firms had leveraged their innovativeness to design the following mechanisms:

(i) 'Formal meetings' involving middle management and employees were organized to share KPIs trends, lessons learned as well as new short and middle term objectives;
(ii) 'Playful activities' were organized on a regular basis to create organizational life and cohesion;
(iii) 'Elaborate posters' were designed and positioned in recreation areas so as to provide information on how the job of workers and managers fitted into the firm's value chain;

(iv) 'Collaborative scenario analysis' and follow-up activities (i.e., joint development of final reports) were organized to promote common goals with supply chain partners and provide opportunities to improve inter-organizational coordination;

(v) Supplier questionnaires, typically seen as basic arm's length instruments to assess and control behaviors, were instead used as an 'external knowledge management tool' to enhance suppliers' commitment to work with (rather than for) the firm. For instance, such questionnaires were having sections allowing suppliers to share visions about future business developments.

Differently from firm E that imposed a strict top-down process to win rigid internal and external structures, Firms I and F managed the deployment of their TESQ strategy as an on-going, cyclical process of continuous improvement and inter-organizational learning. In this path, the mechanisms listed above were of use to generate, share and re-exam information over the way their sustainability strategy should be turned into successful action (Table 2). From this evidence, we propose the following proposition:

P2 As innovativeness increases, the adoption of SSCM practices is more likely to be extensive and successful.

P2a. *As innovativeness increases, a firm's sustainability strategy is more likely to cascade down into successful SPM adoption as the organization employ mechanisms (i.e. formal meetings, playful activities, elaborate posters) that foster employees' commitment and horizontal coordination.*

P2b. *As innovativeness increases, a firm's sustainability strategy is more likely to cascade down into successful SSM adoption as the organization employ mechanisms (i.e. collaborative scenario analysis, external knowledge management tools) that foster suppliers' commitment and vertical coordination.*

5 Discussion

5.1 An Integrative View

Collectively, the overall conceptual model that emerges from our case analysis is depicted in Fig. 1. In line with previous literature, we have found evidence that myopia (Porter and Van der Linde 1995) and complex internal and external structures (Pagell et al. 2013; Walker et al. 2008) impede the development of TESQ strategies and constrain the adoption of SSCM practices. Most importantly, our findings complement prior literature on the relationship between innovation and sustainability (Gualandris and Kalchschmidt 2014; Klassen and Vereecke 2012; van Bommel 2011), providing a refined understanding of how innovativeness fosters SSCM (Fig. 1).

First, firms with high innovativeness were best at anticipating pressure as they created thinking chambers where external stakeholders could share and discuss their

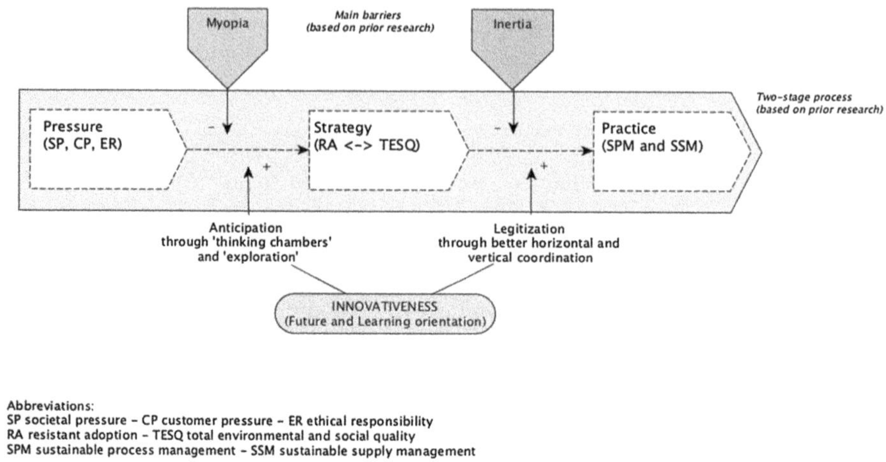

Fig. 1 A conceptual model of how innovativeness fosters SSCM

expectations. We observed that thinking chambers occurred in the form of 'Innovation mall' and 'innovation community'. Such mechanisms allow collecting information and better assessing future sustainability trends; also, they allow evaluating the value different strategic options could potentially deliver to stakeholders as well as the value the firm could ideally retain/capture. Overall, these thinking chambers were observed to enable the development of TESQ strategies. This result is in line with González-Benito and González-Benito (2006), which suggested that the perception of stakeholder pressure and its effect on a firm's environmental strategy is dependent on a firm's managerial attitude. This result also complements prior research suggesting that strategies for which positive private and social benefits coincide are usually the result of a constructive process of interactions between the firm and its prominent network of stakeholders (Pagell and Wu 2009; Schaltegger and Wagner 2011).

Then, we observed that high future orientation associates with mechanisms such as 'idea competitions', 'reimbursements' and 'open meetings'. Such mechanisms were observed to offer concrete opportunities to ethically responsible employees to manifest and pursue their values through concrete explorative projects. Such projects were observed to be economically successful, thus providing evidence of how turning ethical responsibility into strategy could be valuable for the firm. This finding is in line with prior research suggesting that individuals within organization can be entrepreneurs (Schaltegger and Wagner 2011) or champions (Ehrgott et al. 2011); while pursuing their creativity and ambition, such individuals enable and accelerate a firm's change for sustainability. This also complements prior literature suggesting that new forms of work organizations (employee participation) enable the development of advanced sustainability strategies (Longoni 2014).

Next, we observed that firms with high innovativeness were best at legitimizing the adoption of new practices, both internally (SPM) and in their supply chains

(SSM). Prior research suggest that when a new strategy has to be pursued, the violation of taken-for-granted identity codes, ways of thinking, routines or practices usually sets off cascades of radical changes that generally require firms to have sufficient relational expertise (Gualandris et al. 2014; Parmigiani et al. 2011). Relational expertise is accumulated as a firm designs formal and informal mechanisms to share information, increase commitment, and generate common goals within and outside organizations (Calantone et al. 2002; Holcomb and Hitt 2007). Our study complements such studies in suggesting that relational expertise enables innovative firms to develop specific coordination mechanisms ('formal meetings', 'playful activities', 'elaborate posts', 'collaborative scenario analysis' and 'external knowledge management tools') and deploy them along a cyclical process of action and revision. Along such process, employees and suppliers receive incentives and support to accept new operational paradigms, thus facilitating the institutionalization of SSCM.

5.2 Practical Implications

It is increasingly clear that organizations will need to deal with environmental and social issues, reducing their harm and increasing regenerative impacts on social and natural systems (Pagell and Wu 2009). However, turning pressure into strategy, and strategy into practice is never but easy. Changing stakeholders' expectations make it difficult for a firm to understand the boundaries of its responsibility and evaluate the value of alternative courses of actions, thus hampering the process through which objectives and plans are defined. Also, the adoption of new SSCM practices is usually welcomed by employees and supply chain partners with attrition, as changing consolidated routines and ways of operate can be costly and stressful.

Based on our case studies, we recommend managers to open their foresight and improve their learning orientation. Requests and requirements from primary (customer) and societal stakeholders (Government, non-governmental organizations) can be anticipated/influenced by creating thinking chambers and developing strong communication ties. To this end, we suggest firms to design and deploy the following mechanisms: innovation mail and innovation community, i.e., newsletters and blogs which functions as market place for problems, ideas and solutions. Nonetheless, firms are suggested to leverage ethical values and capability sets of champions or intrapreneurs within their organization. Here, mechanisms such as 'Ideas competitions', 'reimbursements' and 'open meetings' naturally provide means for valuable employees to manifest their value, promote innovation and help the firm finding economically-viable strategies to anticipate/meet stakeholders' expectations.

Then, firms must realize that change management based on hierarchical power and top-down approaches does not work well in the context of sustainability. Successful deployments of sustainability strategies are observed when authority is built upon extensive relational expertise. To this end, we suggest firms to experience them-selves with the development of 'formal meetings', 'playful activities' and 'elaborate posters'. Overall, these mechanisms fosters internal cohesion, knowledge sharing, and motivation within the organization. Similarly, a successful adoption of practices that require transactions with suppliers (SSM) is observed when the firm establishes a cyclical process of inter-organizational learning. To this end, 'collaborative scenario analysis' and other 'external knowledge management tools' (i.e., refined questionnaires) should work at best.

Finally, a valuable contribution stems from our assessment framework (Tables 3, 4, 5 and 6). Our analysis has indirectly proved the effectiveness of this tool in measuring external pressure, ethical responsibility, sustainability strategy, SSCM adoption and innovativeness. Thus, we argue that practitioners should purposefully use it along the process through which pressure is turned into practice. First, it can serve as a 'control' tool; managers can indeed evaluate whether pressure, strategy and adoption are coherent. Second, the tool can be used to support 'continuous improvement' and 'communication' with both internal and external stakeholders. As it allows identifying misalignments and lacks of innovativeness, it allows better justification for new investments that could create synergistic effects for innovativeness and sustainability.

5.3 Limitations

Several limitations of this research suggest avenues for future research. First, the relationship between innovativeness and sustainability is here assumed to be mono-directional (e.g. innovativeness leads to higher sustainability), while actually it could be also the other way round (e.g. sustainability leads to higher innovativeness) (Nidumolou et al. 2009). Furthermore, some of the constructs studied in this work (innovativeness, ethical responsibility) are difficult to assess, as they are tacit, unarticulated, shared across a diverse range of organizational functions, and developed over time. Future studies are thus suggested to implement longitudinal analysis to shed more light on how sustainability and innovation evolve and interact over time.

Second, we found indications that innovativeness is necessary to overcome the inertia characterizing complex business settings. Future works could focus on the effect that a misalignment between innovativeness and the complexity of a firm's internal and external structures may have on a firm's environmental and social performance. Third, some of the firms in our sample are part of important business groups. Being part of such networks influences a firm's resources, capabilities and strategies (Chang and Choi 1988), thus likely affecting the degree of innovativeness and its role along the process through which pressure is turned into practice. For instance, recent literature suggests that small and medium enterprises can overcome their limits in pursuing environmental innovation (i.e., limited resources) (Baylis et al. 1998) because of their capability to leverage weak and strong ties with other

Table 3 The role of pressures

Specific dimensions for pressure evaluation	Evaluation guideline		
	1 (Marginal)	2 (Medium)	3 (Relevant)
(a) *Societal pressure SP* (the role of societal stakeholders in inducing/motivating a firm to address environmental and social issues through regulations, laws and withholding/boycotting actions)	–Environmental and social laws, their tightening as well as the threat of withholding/boycotting actions do not motivate a firm's objectives and plan for sustainable development	–Environmental and social laws, their tightening as well as the threat of withholding/boycotting actions are in part responsible for a firm's objectives and plan for sustainable development	–Environmental and social laws, their tightening as well as the threat of withholding/boycotting actions motivate a firm's objectives and plan for sustainable development
1. The extent to which environmental and social laws and legislations already in force are responsible for a firm's objectives and plan for sustainable development			
2. The extent to which the prospective tightening of laws and legislation determines a firm's objectives and plan for sustainable development			
3. The extent to which the threat of withholding and boycotting actions might disrupt the supply chain motivates a firm's objectives and plan for sustainable development			
(b) *Customer pressure CP* (the role of main customers in inducing/motivating a firm to address environmental and social issues through request and requirements)	–Request and requirements by customers to reduce environmental and social impacts of products and operations, and/or to produce detailed information to assure sustainability compliance do not motivate a firm's objectives and plan for sustainable development	–Request and requirements by customers to reduce environmental and social impacts of products and operations, and/or to produce detailed information to assure sustainability compliance are in part responsible for a firm's objectives and plan for sustainable development	–Request and requirements by customers to reduce to reduce environmental and social impacts of products and operations, and/or to produce detailed information to assure sustainability compliance substantially motivate a firm's objectives and plan for sustainable development
1. The extent to which request and requirements by customers to reduce environmental and social impacts of products and processes determine a firm's objectives and plan for sustainable development			

(continued)

Table 3 (continued)

Specific dimensions for pressure evaluation	Evaluation guideline		
	1 (Marginal)	2 (Medium)	3 (Relevant)
2. The extent to which customers' request of detailed information to assure sustainability compliance motivates a firm's objectives and plan for sustainable development			
(c) *Ethical responsibility ER* (the genuine intention and willingness of the top management, middle management and employees in a firm to be engaged in environmental and social management)	–Lack of genuine interest toward sustainability issues by a firm's organizational constituents	–Some individuals (at different organizational level) within a firm believe that environmental and social issues are important	–A firm's organizational constituents strongly believe that caring about environmental and social issues is the right thing to do, thus ethical responsibility of individuals motivate a firm's objectives and plan for sustainable development
1. The extent to which top and middle management in a firm believe that caring about environmental and social issues is 'the right thing to do'			
2. The extent to which a firm's employees express genuine interest toward sustainability issues			

Table 4 Type of sustainability strategy

Specific dimensions for sustainability strategy evaluation	Evaluation guideline		
	1 (Resistant adaptation)	2 (Medium)	3 (TESQ)
Sustainability strategy SS (the set of environmental and social objectives, plans and procedures of a firm)			
1. The extent to which sustainability represent an order winner in a firm business	–A firm sustainability objective consists of complying with legislations and accommodating only salient requests from customers	–A firm sustainability objective is not just limited to complying with legislations, but rather it also attends to the requirements of external stakeholders (e.g., customers)	–The sustainability objective is one of the priority objectives of a firm
2. The amount of time and financial resources a firm allocates to plan and implement sustainability-related actions	–A firm dedicates minimum time and/or financial resources to improve environmental and social performance	–A firm dedicates the necessary time and resources to sustainability performance in order to attend external pressure	–A firm dedicates important budgets to sustainability for reasons that go beyond attending external pressure
3. The extent to which a firm assigns responsibilities for environmental and social matters	–A firm does not have any person who is responsible for dealing with environmental and social matters	–A firm requests the services of external professionals and/or has qualified internal personnel to take care of sustainability matters	–A firm clearly assigns responsibility for sustainability matters to one or various persons of the organization who are specialized in this matter

Table 5 The extent of SSCM adoption

Specific dimensions for SSCM adoption evaluation	Evaluation guideline		
	1 (Low)	2 (Medium)	3 (High)
(a) *Sustainable process management SPM*			
1. Hazardous material management within the product	–None	–Limited consideration of hazardous material monitoring and elimination	–Complete elimination of the current and likely-to-be hazardous materials
2. Use of eco-design (design for environment and life cycle assessment)	–None	–Limited and/or less important consideration of simple criteria regarding sustainability performance in the new product development process	–Complete integration of sustainability performance evaluation and monitoring procedures into the new product development process
3. Building an environmental management system (EMS)	–None	–EMS implementation in some managerial processes (e.g., operations) and obtained environmental certification	–Completed implementation of an EMS throughout the whole organization
4. Reduction of energy, water use and emission	–None	–Energy and emission reduction practices in few operations	–'reduction at the source' practices embedded in every operation
5. Improving health and safety (e.g., obtaining OHSAS 18001)	–None	–Health and safety improvement practices in few operations	–Health and safety improvement practices in the full-range of firm's operations
6. Improving corporate responsibility through social campaigns (e.g., codes of conduct, corporate social activities)	–None	–Some effort is put in the development of a firm code of conducts/ethics	–Established code of conduct/ethics

(continued)

Table 5 (continued)

Specific dimensions for SSCM adoption evaluation	Evaluation guideline		
	1 (Low)	2 (Medium)	3 (High)
(b) *Sustainable supply management SSM (supplier assessment and collaboration)*			
(1a) Greener part and material provide (e.g., harm-free or more recyclable products)	–A firm hardly requires or considers those factors when selecting or evaluating suppliers	–A firm officially requires its suppliers to improve their performance in those elements	–A firm demands its suppliers to undertake environmental and social practices very seriously. Also, the portion of environmental and social criteria in a firm's procurement process is very substantial
(1b) A firm-wide environmental management system or certification			
(1c) A firm effort to improve health and safety in operations and behave ethically			
(2a) Diagnosis and improvement of the environmental and social profile of supplier operations	–A firm does not provide any managerial or technical assistance, to its suppliers or even work with them to implement such practices	–A firm works with suppliers to implement such practices, but there is a lack of structured procedures and the collaboration is limited to a small portion of suppliers	–A firm has clear procedure and extensively works with a substantial number of suppliers for improving environmental and social performance
(2b) Environmental and social certifications with suppliers			
(2c) Collaborative greener product development			

Table 6 Degree of Innovativeness

Specific dimensions for innovativeness evaluation	Evaluation guideline		
	1 (Low)	2 (Medium)	3 (High)
(a) *Future orientation FO*			
1. Dominant paradigm	–A firm believes that the future can be known by means of expertise	–A firm believes that the future can be known by means of projections and calculations	–A firm believes that the future can be shaped by means of interaction with the business surrounding environment
2. Perspective	–Exploring change	–Reacting to change	–Understanding and anticipating/shaping change
3. Foresight process	–Collect and compare the opinions of experts	–Use quantitative models to estimate the future (e.g., S-curve)	–Open and continuous dialogue to leverage sense-making and sense-giving processes
4. Outsiders	–Outsiders (experts) are responsible for both contents and outcomes of the foresight process	–Some of the foresight activities are outsourced to outsiders which generate foresight knowledge	–Strong collaboration with all relevant stakeholders (in and out-side)
5. Output activities	–Scarce investments in building roadmaps, scenarios and Delphi	–Trend-databases, matrices, monitoring systems	–Scenarios, innovation ideas, wild card, action plans
(b) *Learning orientation LO* (*commitment to learning, shared-vision and memory*)			
(1a) Perceived importance of the organization's ability to learn (1b) The sense associated to continuous training and learning by organizational constituents	–A firm perceives continuous learning as an expense	–Limited consideration of continuous learning processes	–A firm perceives learning as a key investment (not an expense) to guarantee organizational survival and competitive advantage
(2a) Degree of commonality of purposes (2b) Degree of agreement on the organizational vision (2c) Degree of employees commitment to organizational goals	–A firm misses efficient and effective mechanisms to share and consolidate its vision throughout the organization	–There is communality of purpose and intent in a firm	–Organizational constituents across levels, functions and divisions are committed to the firm's vision and have a clear understand of how their work fits into the value chain of the firm

(continued)

Table 6 (continued)

Specific dimensions for innovativeness evaluation	Evaluation guideline		
	1 (Low)	2 (Medium)	3 (High)
(3a) Degree of organizational conversation (3b) Degree of organizational transparency on unsuccessful endeavors (3c) Degree to which lessons learned are shared throughout the whole organization	–A firm never analyzes unsuccessful endeavors and its members do not stress the distribution and preservation of knowledge	–Some mechanisms for sharing and reexamining information are developed which allow accumulating knowledge	–A firm always analyzes unsuccessful endeavors, has specific mechanisms for sharing lessons learned and maintains organizational conversation to keep alive such lessons from history

affiliated firms (Halila 2007). An interesting research avenue, thus, would be to study the relationship between business group affiliation, innovativeness and sustainability.

Finally, concerning generalizability, our findings and considerations are based on case research with only Italian firms operating in manufacturing sectors. The role and effectiveness of innovativeness and its underlying mechanisms, however, might vary from country to country since cultural settings matter; a capability for anticipation, for instance, is not equally important in all cultures (Kirkman et al. 2006). Thus, future research should study if and how cultural differences affect the role of innovativeness for sustainability.

6 Conclusions

This research was aimed at deepening the understanding of how and through which mechanisms innovativeness assists a firm in turning stakeholders' pressure into sustainability strategy, and then strategy into successful SSCM adoption. An assessment framework was built and deployed in a continuous process of theory building and data collection through case studies. A set of proposition and a conceptual model were finally developed that shed light on the relationship between innovativeness and sustainability. Findings show that, as innovativeness increases, firms are more capable at (i) anticipation or capturing and leveraging weak signals from within and outside the organization to anticipate/shape external pressure, and (ii) legitimization or sharing and re-examining information over the deployment of sustainability objectives and plans so to overcome the inertia. Anticipation is enabled by tools and routines such as 'Innovation mall', 'innovation community', 'idea competitions', 'reimbursements' and 'open meetings', which allow to create thinking chambers with primary and societal stakeholders as well as leverage employees' values and capabilities. Legitimization is enabled by tools and routines such as 'formal meetings', 'playful activities', 'elaborate posts', 'collaborative scenario analysis' and 'external knowledge management tools', which foster commitment within and outside the firm and enact better horizontal/vertical coordination. Overall, the outcomes of this study complement prior literature suggest future research avenues and provide guidance to firms that struggle with how to manage pressure and deliver new sustainability standards in their supply chain.

References

Ageron B, Gunasekaran A, Spalanzani A (2012) Sustainable supply management: an empirical study. Int J Prod Econ 140:168–182

Ahi P, Searcy C (2013) A comparative literature analysis of definitions for green and sustainable supply chain management. J Cleaner Prod 52:329–341

Ashford NA, Hall RP, Ashford RH (2012) The crisis in employment and consumer demand: reconciliation with environmental sustainability. Environ Innov Societal Transitions 2:1–22
Ateş MA, Bloemhof J, van Raaij EM, Wynstra F (2011) Proactive environmental strategy in a supply chain context: the mediating role of investments. Int J Prod Res 50:1079–1095
Awaysheh A, Klassen RD (2010) The impact of supply chain structure on the use of supplier socially responsible practices. Int J Oper Prod Manage 30:1246–1268
Bansal P, Roth K (2000) Why companies go green: a model of ecological responsiveness. Acad Manage J 43:717–736
Baylis R, Connell L, Flynn A (1998) Company size, environmental regulation and ecological modernization: further analysis at the level of the firm. Bus Strateg Environ 7:285–296
Bocken NMP, Short SW, Rana P, Evans S (2014) A literature and practice review to develop sustainable business model archetypes. J Cleaner Prod 65:42–56
Bos-Brouwers HEJ (2010) Corporate sustainability and innovation in SMEs: evidence of themes and activities in practice. Bus Strateg Environ 19:417–435
Calantone RJ, Cavusgil ST, Zhao Y (2002) Learning orientation, firm innovation capability, and firm performance. Ind Mark Manage 31:515–524
Carroll AB (1999) Corporate social responsibility. Bus Soc 38:268
Carter CR, Jennings MM (2004) The role of purchasing in corporate social responsibility: a structural equation analysis. J Bus Logistics 25:145–186
Chang SJ, Choi U (1988) Strategy, structure and performance of Korean business groups: a transactions cost approach. J Ind Econ 37:141–158
Christmann P (2000) Effects of "best practices" of environmental management on cost advantage: the role of complementary assets. Acad Manage J 43:663–680
Christmann P (2004) Multinational companies and the natural environment: determinants of global environmental policy standardization. Acad Manage J 47:747–760
Daily BF, Huang S (2001) Achieving sustainability through attention to human resource factors in environmental management. Int J Oper Prod Manage 21:1539–1552
Darnall N, Jolley GJ, Handfield R (2008) Environmental management systems and green supply chain management: complements for sustainability? Bus Strateg Environ 17:30–45
Ehrgott M, Reimann F, Kaufmann L, Carter CR (2011) Social sustainability in selecting emerging economy suppliers. J Bus Ethics 98:99–119
Eisenhardt KM, Graebner ME (2007) Theory building from cases: opportunities and challenges. Acad Manage J Arch 50:25–32
Eisenhardt KM, Martin JA (2000) Dynamic capabilities: what are they? Strateg Manage J 21:1105–1121
Garcia R, Calantone R (2002) A critical look at technological innovation typology and innovativeness terminology: a literature review. J Prod Innov Manage 19:110–132
González-Benito J, González-Benito Ó (2006) A review of determinant factors of environmental proactivity. Bus Strateg Environ 15:87–102
Gualandris J, Golini R, Kalchschmidt M (2014) Do supply management and global sourcing matter for firm sustainability performance? An international study. Supply Chain Manage Int J 19:258–274
Gualandris J, Kalchschmidt M (2014) Customer pressure and innovativeness: their role in sustainable supply chain management. J Purchasing Supply Manage 20:92–103
Halila F (2007) Networks as a means of supporting the adoption of organizational innovations in SMEs: the case of environmental management systems (EMSs) based on ISO 14001. Corp Soc Responsib Environ Manage 14:167–181
Handfield RB, Walton SV, Seegers LK, Melnyk SA (1997) 'Green'value chain practices in the furniture industry. J Oper Manage 15:293–315
Herring H, Sorrell S (2009) Energy efficiency and sustainable consumption: the rebound effect. Palgrave Macmillan, Basingstoke
Holcomb TR, Hitt MA (2007) Toward a model of strategic outsourcing. J Oper Manage 25:464–481
Hult GTM, Ketchen DJ (2003) Organizational learning as a strategic resource in supply management. J Oper Manage 21:541–556

Isaksson R, Johansson P, Fischer K (2010) Detecting supply chain innovation potential for sustainable development. J Bus Ethics 97:425–442

Kirkman BL, Lowe KB, Gibson CB (2006) A quarter century of culture's consequences: a review of empirical research incorporating Hofstede's cultural values framework. J Int Bus Stud 37:285–320

Klassen RD (2001) Plant-level environmental management orientation: the influence of management view and plant characteristics. Prod Oper Manage 10:257–275

Klassen RD, Vachon S (2003) Collaboration and evaluation in the supply chain: the impact on plant-level environmental investment. Prod Oper Manage 12:336–352

Klassen RD, Vereecke A (2012) Social issues in supply chains: capabilities link responsibility, risk (opportunity), and performance. Int J Prod Econ 140:103–115

Krause DR, Vachon S, Klassen RD (2009) Special topic forum on sustainable supply chain management: introduction and reflections on the role of purchasing management. J Supply Chain Manag 45:18–25

Lee SY, Klassen RD (2009) Drivers and enablers that foster environmental management capabilities in small-and medium-sized suppliers in supply chains. Prod Oper Manage 17:573–586

Lee SY, Rhee SK (2007) The change in corporate environmental strategies: a longitudinal empirical study. Manage Decis 45:196–216

Longoni A (2014) HRM and organisational practices in operations: the impact on environmental and social sustainability, sustainable operations strategies. Springer, New York, pp 13–20

Maignan I, Hillebrand B, McAlister D (2002) Managing socially-responsible buying: how to integrate non-economic criteria into the purchasing process. Eur Manage J 20:641–648

Murillo-Luna JL, Garcés-Ayerbe C, Rivera-Torres P (2008) Why do patterns of environmental response differ? A stakeholders' pressure approach. Strateg Manage J 29:1225–1240

Nidumolu R, Prahalad CK, Rangaswami M (2009) Why sustainability is now the key driver of innovation. Harvard Bus Rev 87:56–64

Pagell M, Gobeli D (2009) How plant managers' experiences and attitudes toward sustainability relate to operational performance. Prod Oper Manage 18:278–299

Pagell M, Johnston D, Veltri A, Klassen R, Biehl M (2013) Is safe production an oxymoron? Prod Oper Manage 23:1161–1175

Pagell M, Shevchenko A (2014) Why research in sustainable supply chain management should have no future. J Supply Chain Manage 50:44–55

Pagell M, Wu Z (2009) Building a more complete theory of sustainable supply chain management using case studies of 10 exemplars. J Supply Chain Manage 45:37–56

Parmigiani A, Klassen RD, Russo MV (2011) Efficiency meets accountability: performance implications of supply chain configuration, control, and capabilities. J Oper Manage 29:212–223

Porter ME, Van der Linde C (1995) Green and competitive: ending the stalemate. Harvard Bus Rev 73:120–134

Reuter C, Foerstl K, Hartmann E, Blome C (2010) Sustainable global supplier management: the role of dynamic capabilities in achieving competitive advantage. J Supply Chain Manage 46:45–63

Robson LS, Clarke JA, Cullen K, Bielecky A, Severin C, Bigelow PL, Irvin E, Culyer A, Mahood Q (2007) The effectiveness of occupational health and safety management system interventions: a systematic review. Saf Sci 45:329–353

Rohrbeck R, Gemünden HG (2011) Corporate foresight: its three roles in enhancing the innovation capacity of a firm. Technol Forecast Soc Change 78:231–243

Ruff F (2006) Corporate foresight: integrating the future business environment into innovation and strategy. Int J Technol Manage 34:278–295

Schaltegger S, Wagner M (2011) Sustainable entrepreneurship and sustainability innovation: categories and interactions. Bus Strateg Environ 20:222–237

Schumpeter JA (1934) The theory of economic development: an inquiry into profits, capital, credit, interest, and the business cycle. Transaction Publishers, New Jersey

Seuring S, Muller M (2008) From a literature review to a conceptual framework for sustainable supply chain management. J Cleaner Prod 16:1699–1710

Sharma S, Henriques I (2005) Stakeholder influences on sustainability practices in the Canadian forest products industry. Strateg Manage J 26:159–180

Sinkula JM, Baker WE, Noordewier T (1997) A framework for market-based organizational learning: linking values, knowledge, and behavior. J Acad Mark Sci 25:305–318

Stuart I, McCutcheon D, Handfield R, McLachlin R, Samson D (2002) Effective case research in operations management: a process perspective. J Oper Manage 20:419–433

United Nation Global Compact and Business for Social Responsibility (2010) Supply chain sustainability. A practical guide for continuos improvement. UN Global Compact Office

van Bommel HWM (2011) A conceptual framework for analyzing sustainability strategies in industrial supply networks from an innovation perspective. J Cleaner Prod 19:895–904

van Tulder R, van Wijk J, Kolk A (2009) From chain liability to chain responsibility. J Bus Ethics 85:399–412

Verona G (1999) A resource-based view of product development. Acad Manage Rev 24:132–142

Walker H, Di Sisto L, McBain D (2008) Drivers and barriers to environmental supply chain management practices: lessons from the public and private sectors. J Purchasing Supply Manage 14:69–85

Wu Z, Pagell M (2011) Balancing priorities: decision-making in sustainable supply chain management. J Oper Manage 29:577–590

Yin RK (2009) Case study research: design and methods. Sage Publications, Inc., New York

Zairi M, Peters J (2002) The impact of social responsibility on business performance. Manag Auditing J 17:174–178

Zhu Q, Sarkis J (2004) Relationships between operational practices and performance among early adopters of green supply chain management practices in Chinese manufacturing enterprises. J Oper Manage 22:265–289

Zhu Q, Sarkis J, Geng Y (2005) Green supply chain management in China: pressures, practices and performance. Int J Oper Prod Manage 25:449–468

Zhu Q, Sarkis J, Lai K (2007) Green supply chain management: pressures, practices and performance within the Chinese automobile industry. J Cleaner Prod 15:1041–1052

Zhu Q, Sarkis J, Lai K (2012) Examining the effects of green supply chain management practices and their mediations on performance improvements. Int J Prod Res 50:1377–1394

Sustainable Manufacturing: The Lean and Green Business Model

Andrea Brasco Pampanelli, Pauline Found and Andréa Moura Bernardes

Abstract Over the past few years there has been an emergent trend towards integrating a lean and green approach with a number of papers and books written, but most of these take an outside-in approach describing case studies from industry. This paper differs in that it takes an inside-out approach; describing a Lean and Green Business Model (L&GBM) developed within a major global engineering company. It describes the five key principles of a Lean and Green Business Model, (i) a stable value stream, (ii) identification of environmental impacts, (iii) measurement the environmental value streams, (iv) improvement of the environmental value streams and (v) continuous improvement. It further explains how the model applies a Kaizen approach for improving mass and energy flows of manufacturing environment that already possesses a basic deployment level in applying lean. Some of the key findings identified by the researchers highlight that (i) L&GBM has a different purpose than traditional Lean or Environmental Thinking, (ii) L&GBM covers the three dimensions of sustainability, (iii) L&GBM has a Lean to Green approach and (iv) L&GBM is an alternative approach to integrate environmental concerns into operations management which enhances workplace engagement in reducing the environmental impact of the manufacturing processes by leveraging the lean attributes of involvement and empowerment to the environmental functions within the organization that traditionally have focused on compliance.

A.B. Pampanelli
GKN Driveline Americas, Rua Joaquim Silveira 557,
91060-320 Porto Alegre, Brazil
e-mail: andrea.pampanelli@gkndriveline.com

P. Found (✉)
Buckingham Business School, University of Buckingham,
Buckingham MK18 1EG, UK
e-mail: pauline.found@buckingham.ac.uk

A.M. Bernardes
UFRGS, PPGEM Doctoral Programme, Av. Bento Gonçalves,
9500, Setor 4, Campus do Vale, 91501-970 Porto Alegre, Brazil
e-mail: amb@ufrgs.br

A. Chiarini (ed.), *Sustainable Operations Management*,
Measuring Operations Performance, DOI 10.1007/978-3-319-14002-5_7

Keywords Lean · Sustainability · Lean and green · Kaizen

1 Introduction and Purpose

Sustainability has become a legacy for the 21st century. It embodies the promise of societal evolution towards a more equitable and richer world in which the natural environment is preserved for generations to come. The quest for economic growth and social equity has become a major goal for most of the past 150 years. By adding concern for the carrying capacity of natural systems, sustainability ties together the current main challenges facing humanity.

Although the issues embodying sustainability are more than a century old, the concept of sustainable development itself was first described in the late 80s, following The Brundtland Report, a report by the World Commission on Environment and Development (WCED 1987) that describes the growing global awareness of the enormous environmental problems facing the planet, and proposes a shift towards global environmental action. The concern about sustainability encouraged society to support the development of a significant number of corporate practices, many applied to manufacturing business, such as Industrial Ecology, Industrial Symbiosis, Pollution Prevention, Cleaner Production, etc. with the ultimate goal of the supporting the sustainability dimensions of (1) profit, (2) people and (3) planet or the triple bottom line (Elkington 1997). Although all these studies and practices have contributed to create a new world paradigm, very few were able to contribute fully to all dimensions of sustainability (Lozano 2012). The term 'Green' is used in this paper and for the model proposed here to cover all concerns for the environmental impact of manufacturing and, in particular, to address the planet dimension of sustainability.

...manufacturing is the constant game of doing more with less... therefore manufacturing managers are constantly looking for new approaches to increase efficiency (Hopp and Spearman 2008). With the purpose of promoting a continuous improvement culture within the business, the expenditure of resources for any goal, other than the creation of value for the end customer, is considered to be wasteful. Lean thinking is one of these strategies that are explored by manufacturing to increase performance, contributing to the profit dimension, by developing and respecting people. The logic of lean thinking, with the emphasis on eliminating the seven classic wastes (Ohno 1988) can be redesigned and integrated to include an environmental, or green, dimension of sustainability, addressing all three dimensions of profit, planet and people.

...a gram of prevention is better than a kilogram of cure... therefore using less energy, material, generating less waste is prevention, and so good for the environment (Baas 2007). Minimizing waste produced in manufacturing, reducing the energy use and using the materials and resources in a more efficient way can lead to financial cost savings and a reduction of environmental impacts. Therefore, integrating both concepts, lean thinking and sustainability, offers the foundation for a new business logic, where the pillars of sustainability, social, economic and environmental, can be understood by manufacturing and therefore supports business goals, requirements and needs.

The main objective of this paper is to propose a Lean and Green Business Model (L&GBM) where the environmental aspect of sustainability is added to the pure lean thinking concept in order to create a way of thinking that contributes to, and balances, the three sustainability dimensions of people, profit and planet (Elkington 1997). This model takes the Kaizen (continuous improvement) approach for improving mass and energy flows in a manufacturing environment that already possesses a deployment level in applying lean. Figure 1 presents the main role of the study.

This paper is based on action research developed from 2009 to 2013 by a team of lean and environmental experts from Brazil and UK. As an overall objective, it aims to propose a new and integrated way of thinking that:

- Integrates the pure 'Lean Thinking' concepts with an environmental, or 'Green Thinking' dimension;
- Contributes to, and balances, the three sustainability dimensions, or triple bottom-line of people (social sustainability) profit (economic sustainability) and planet (environmental sustainability);
- Uses the Kaizen approach for managing and improving environmental flows of mass and energy in manufacturing environment.

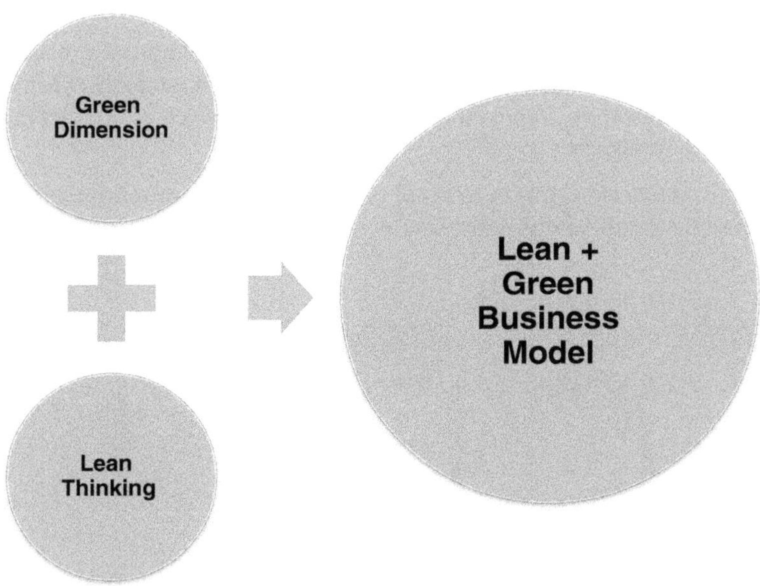

Fig. 1 The main objective of the study

To contextualize the subject and support the problem definition, this paper will be answering the following research questions:

- Can lean manufacturing practices be adapted and used as a strategy to achieve business environmental sustainability?
- What different frameworks and corporate strategies are needed to support the three sustainability dimensions?

In order to create the basis for the L&GBM, this paper explores some of the fundamental building blocks of lean thinking with sustainability and green concepts. It proposes the model structure and dynamics. The paper also reports the structure of the new model, considering purpose, principles and ways of working, discussing why it is different from pure green and pure lean thinking.

Action research or participatory action research is a reflective process of progressive problem solving led by individuals working with others in teams, or as part of a community of practice, to improve the way they address issues and solve problems. Action research involves the process of actively participating in an organization change situation whilst conducting research. Action research can also be undertaken by larger organizations or institutions, assisted or guided by professional researchers, with the aim of improving their strategies, practices, and knowledge of the environments within which they practice. As designers and stakeholders, researchers work with others to propose a new course of action to help their community improve its work practices. According to Gill (2009) an action research type of study is developed in seven steps: (1) exploratory phase, (2) understanding the problem, (3) defining the hypothesis, (4) project scope, (5) data collection and (6) analysis of results. Figure 2 presents the basic framework for the research structure.

The analysis of the key findings and improvement opportunities related to the application of the L&GBM were developed in five different moments. Each of these moments was called an improvement cycle. The objectives of developing these improvement cycles are:

- Analysis of kaizen results in terms of (1) reduction of environmental impact, (2) increase the productivity in the use of resources;

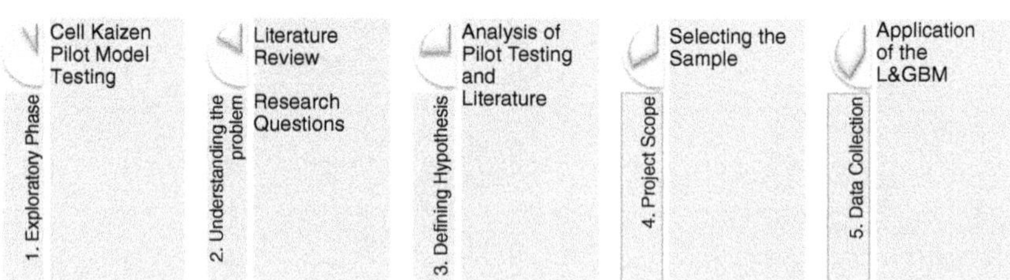

Fig. 2 Research structure applied for the development of the L&GBM. *Source* Developed by the authors

- Analysis of action plan results in terms of cost reduction;
- Confirm model prerequisites;
- Identify other key findings;
- Identify model improvement opportunities.

In this study the L&GBM model was developed, tested and improved through a series of iterative testing cycles and the analysis of the results of each cycle. Two methods were applied for developing these analyses: (1) Brainstorm sessions with participants and specialists and (2) A3 analysis. Brainstorming is a group creativity technique by which a group tries to find a solution to a specific problem by gathering a list of ideas spontaneously contributed to by its members. Brainstorming was developed and coined by Osborn (1963) through the book Applied Imagination. A3 is a structured problem-solving approach developed by Toyota for training of engineers, supervisors and managers. The term A3 derives from the paper size used for the report, which is the metric equivalent to 11 × 17 paper. Toyota actually uses several styles of A3 reports: for solving problems, for reporting project status, and for proposing policy changes. The A3 process helps people engage in collaborative, in-depth problem solving. It drives problem-solvers to addressing the root causes of problems, which surface in day-to-day work routines (Sobek II and Smalley 2008).

2 Problem Definition

In order to contextualize the subject and support the problem definition, this paper discusses the fundamental building blocks of the research, such as the (2.1) *the fundamental aspects of lean thinking,* (2.2) *the basis for green thinking and* (2.3) *the integration of lean and green to develop a model for sustainable manufacturing.* This paper forwards the following propositions:

- The application of pure lean promotes environmental improvement even though there is no direct intention to reduce environmental impact.
- There are several examples describing the synergy between lean and green practices but none, so far, have proposed a different way of thinking.
- Most frameworks and corporate strategies do not fully contribute to the core sustainability dimensions.
- Pure lean thinking contributes to two sustainability dimensions—(1) profit and (2) people.
- A new way of thinking can be created by integrating one further dimension—(3) the planet to pure lean thinking.

2.1 The Fundamental Aspects of Lean Thinking

Manufacturing is the constant game of doing more with less (Hopp and Spearman 2008). Therefore, lean thinking is one of the improvement strategies that have completely changed the way manufacturing has developed over the past decades. Gordon (2001) states that for decades, lean manufacturing has been considered the best way to run a manufacturing company.

According to Bicheno (2000) the general purpose of lean thinking can be described in three main dimensions (1) Quality, (2) Delivery and (3) Cost. It means that, producing exactly what the customer wants, exactly when (with no delay) at fair price and with minimum waste is the ultimate goal of a lean enterprise. Therefore, lean thinking focuses on the optimization of production resources oriented by the customer—time, people, machine, space, etc., and consequently reduces wastes. Thus, Lean Thinking contributes to the economic, or profit, dimension of sustainability. In general terms, lean thinking is defined and described by five key principles (Womack and Jones 1996):

- *Specific value*: define value precisely from the perspective of the end customer in terms of the specific product with specific capabilities offered at a specific time;
- *Identify value streams*: identify the entire value stream for each product or product family and eliminate waste;
- *Make value flow*: make the remaining value creating steps flow;
- *Let the customer pull value*: design and provide what the customer wants only when the customer wants it;
- *Pursue perfection*: strive for perfection by continually removing successive layers of waste as they are uncovered.

Resource productivity and closed loops provide better services, for longer periods, with less material, cost and hassle. The logic of lean thinking, with the emphasis on eliminating the classic seven wastes (Ohno 1988), makes a customer-defined value flow continuously with the aim of producing less waste. Together these practices offer the foundation for a powerful new business logic: Instead of simply selling the customer a product, it is perceived appropriate, to derive what is desired, considering the quantity, rate and timeliness. Based on the analysis of customer value, lean presents a set of tools and techniques for continuously improving processes and eliminating wastes (Rother and Shook 2003).

Due to the relentless drive to reduce all forms of waste, including defects, over-processing and unnecessary transportation lean contributes to environmental, or green thinking, inadvertently and this paper supports the proposition that:

- *The application of pure lean promotes environmental improvement even though there is no direct intention to reduce environmental impact.*

Table 1 Two major Kaizen objectives

Objectives	Description
Develop a problem solving culture	With a focus on analysis and problem solving by applying scientific and structured thinking. Lean philosophy presents a variety of tools and techniques with the ultimate goal of improving processes and eliminating wastes. Developing a problem solving culture is key for deploying the lean thinking (Berger 1997)
People involvement	Kaizen relies on ongoing effort and engagement of people - it is based on the constant effort for involving and integrates people, from the shop-floor workers to the senior executives. For lean thinking the key for success is based on the capacity for training and involving everyone. Based on this idea, human-systems are considered more successful than software systems for sustaining the results. This creates a learning environment, with long term maintenance of results and openness for creativity and improvements (Berger 1997)

According to Womack and Jones (1996) one of the key building blocks of lean thinking is Kaizen—a process-oriented philosophy that focuses on incremental improvements and standardization of the improved system as the building block for further improvement. Table 1 describes the two major objectives of the Kaizen (Berger 1997):

Whilst Bicheno (2000) considers that lean is described in the QCD dimensions, Hines et al. (2004) argues that pure lean thinking not only focuses in one dimension of sustainability, (1) profit, but also supports another , the (2) people dimension. Considering scientific methods and the involvement of people as the basis for its tools and techniques, lean presents a robust methodology for incorporating the social, people dimension in a system thinking approach. In addition, 'Respect for People' is a key concept of TPS (Sugimori et al. 1977) suggesting that the well being of employees and their involvement in the process improvements is also central to lean. The soft issues of lean, which links the importance of people to the ability to sustain long-term competitive advantages, has been demonstrated in a number of studies (Beale and Found 2006; Carleysmith et al. 2009; Found et al. 2006; Lander and Liker 2007; Liker 2004; Liker and Meier 2007; Liker and Hoseus 2008; Liker and Convis 2012; Mann 2005). Therefore, according to the authors, pure lean thinking contributes to two dimensions of the sustainability concept, such as:

Full contribution to the profit dimension due to its core focus in eliminating the seven classic wastes and reducing costs and;

Partial contribution to the people dimension, due to its focus on the Kaizen continuous improvement philosophy for solving problems and involving people, therefore this paper proposes that:

- *Pure Lean Thinking contributes to two sustainability dimensions*—(1) *profit and* (2) *people.*

2.2 Basis for Green Thinking

Green Thinking is rooted in sustainability, which is a systemic concept relating to the continuity of economic, social and environmental aspects of human society. It is however part of a wider and evolving field of corporate social and environmental responsibility, which in modern times has its origins in Rachel Carson's Silent Spring (Carson 1965) and the Club of Rome's 'Limits to Growth' analysis (Meadows et al. 1972). The term was first used in 1987 by the Brundtland Commission, which coined what has become the most often-quoted definition of sustainable development:

Sustainable development is development that meets the needs of the present without compromising the ability of future generations to meet their own needs (WCED 1987).

The field of sustainable development can be conceptually broken into three constituent parts: environmental sustainability, economic sustainability and socio-political sustainability. Figure 3 presents a representative scheme of sustainable development vectors.

Sustainable development ties together concern for the carrying capacity of natural systems with the social challenges facing humanity (Zokaei et al. 2010). Therefore it contains two key concepts: (1) the concept of needs, in particular the essential needs of the world's poor, to whom overriding priority should be given; (2) the idea of limitations imposed by the state of technology and social organization on the environment's ability to meet present and future needs. All these definitions of sustainable development propose understanding the world as a system —a system that connects space and time. Therefore, the concept of sustainable development is rooted in systems thinking (Lovelock 1990).

Sustainability is a big umbrella term addressing a host of issues (Hall 2010) not all of which are limited to environmental, or 'Green' issues. Many writers emphasize only a few aspects of sustaining the planet in a condition to support life, but the scope of concerns is so broad that it's difficult—or impossible—to think

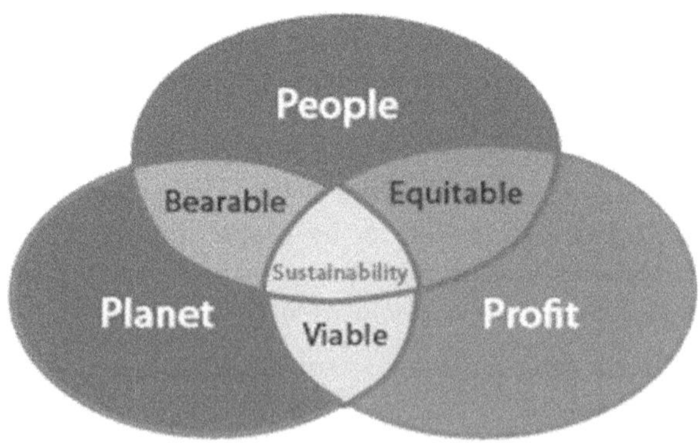

Fig. 3 Sustainable development vectors. *Source* Adapted from Elkington (1997)

about them all at the same time. Consequently, an abundance of separate initiatives attack some aspect of sustainability; local recycling, alternative energy ventures, permaculture, green building codes, etc.

2.2.1 Dimensions of Green Thinking

Over the past decades, many different corporate strategies were created proposing the co-existence of industry, the business, the people, the natural environment and their interactions in systems thinking approach. Zokaei et al. (2010) provides an overview of some of these key management strategies proposed to pursue sustainable development, such as Industrial Ecology (Nielsen 2007; Tibbs 1992), Industrial Symbiosis (Boons et al. 2011), Eco-efficiency (Korhone 2007), Triple Bottom Line (Elkington 1997; Lenzen 2008), Natural Capitalism (Robèrt 2002a; Hawken et al. 1999), The Natural Step (Robèrt 2002b).

In fact, the difficulty to make the concept of sustainability and its application clear is such that many researchers have explored it deeper. Glavic and Lukman (2007) present a study that summarizes the definition of sustainability and its terms. The Stern Review (Stern 2007) explored the economics of climate change. Lozano (2008) identifies the need for many to fully understand the concept, presenting a study that not only expand the concept of sustainability but also clarifies its dimensions. In a second study Lozano (2012) presents research that discusses how company's voluntary sustainability initiatives contribute to the sustainability dimensions. Table 2, adapted from Lozano's study, presents a list of these sustainability/corporate strategies and how they contribute to the sustainability three core dimensions of (1) Economic (Profit) (2) Environment (Planet) and (3) Social (People). With different structure and priorities, all these strategies describe conditions for sustainable systems and propose strategies in order to make sustainable development concept viable.

The conclusion on analyzing Table 2 is that, although all these concepts have been around for much of the last 20 years, most of the sustainability/corporate strategies have focused mainly on environmental conservation and compliance. The majority, as they are proposed, are not integrated, or part of, the fundamental building blocks of the manufacturing strategies that are pro-active in all dimensions of sustainability.

Therefore, following this idea it is possible to conclude that:

- *Lack of existence sustainability/corporate strategies that are able to contribute to the core three dimensions of sustainability (people, profit and planet) and that are fully integrated to the main aspects of the business.*

Table 2 Examples of sustainability/corporate strategies and its contribution to the sustainability dimensions

Sustainability/corporate strategies	Sustainability dimensions		
	Profit/economic	Planet/ environment	People/social
Sustainable livelihoods	*Full contribution*	*Full contribution*	*Full contribution*
Triple bottom line	*Full contribution*	*Full contribution*	*Full contribution*
The natural step	Partial contribution	*Full contribution*	Partial contribution
Environmental management system	None	*Full contribution*	None
Environmental and social accounting	*Full contribution*	*Full contribution*	*Full contribution*
Life cycle analysis	None	*Full contribution*	None
Cleaner production	*Full contribution*	*Full contribution*	None
Design for environment	None	*Full contribution*	None
Eco-efficiency	*Full contribution*	*Full contribution*	None
Industrial ecology	*Full contribution*	*Full contribution*	None
Factor X	Partial contribution	*Full contribution*	None
Green chemistry	None	*Full contribution*	None
Corporate social responsibility	None	Partial contribution	*Full contribution*
Sustainable reporting	*Full contribution*	Partial contribution	Partial contribution
Corporate citizenship	None	None	*Full contribution*

Source Lozano (2012)

2.3 Integrating Lean and Green Thinking

Lean sees waste as anything that is non-value added to the customer (Bicheno 2000). In the other hand, Green sees waste as extraction and consequential disposal of resources at rates, or in forms, beyond that which nature can absorb (Lozano 2008). An environmental waste is an unnecessary, or excessive, use of resources or substances released to the air, water, or land that could harm human health or the environment (EPA 2006). Environmental waste can occur when the company uses resources to provide products or services to customers and/or when customers use and dispose of products (EPA 2006).

There has been debate in the literature whether improving environmental performance would undermine the economic sustainability of an organization and that many businesses could not afford the cost of meeting their environmental responsibilities (Florida 1996; Found 2009). However, there are many examples where improving environmental performance has improved the company's profit (Maxwell et al. 1993; Porter and van der Linde 1995; King and Lenox 2001; Cobert and Klassen 2006; Yang et al. 2011).

The US Environmental Protection Agency (EPA 2006) developed this theme and has reported some key findings: (1) Lean produces an operational and cultural environment that is highly conductive to waste minimization and pollution prevention; (2) Lean can be leveraged to produce even more environmental improvement; (3) Some regulatory friction can be encountered when applying lean to environmentally-sensitive processes; (4) Environmental agencies have a window of opportunity—while companies are embarking on lean initiatives and investments —to collaborate with lean promoters to further improve the environmental benefits associated with lean.

Two recent studies discuss the synergies between pure lean thinking and environmental improvement practices. In the first, Biggs (2009) focused on the integration of lean thinking and environmental improvement. Some of the most important findings she reported were:

- Traditional approaches to lean is capable of providing environmental benefits even though there is no direct intention to reduce environmental impact;
- The lean methodology can be used to make environmental improvements as well as productivity improvements;
- Kaizen/Continuous Improvement (CI) Kaizen blitz and workforce involvement and suggestions are popularly suggested methods of gaining environmental benefit from a Lean implementation;
- It is the culture of waste elimination and experimentation, problem solving and improvement of best practice encouraged by lean that may help companies make environmental improvements;
- A lean approach can help make the business case for environmental impact reduction.

In a second study, Dues et al. (2012) discuss how lean practices are catalysts for greening operations. The authors discuss that the lean and green connection goes beyond the idea of waste reduction, overlapping in paradigms such as (1) tools and practices, (2) supply chain relationship, (3) lead time reduction, (4) focus on people and organization (5) use of techniques for waste reduction. The research findings indicate that green comes as a natural extension to lean as most of lean practices are green without the explicit intention to be green and concludes that lean manufacturers are greener than non-lean companies.

Following these two studies is possible to conclude that:

- *Lean thinking serves as a catalyst to green thinking.*
- *Lean can be the first stage for a company to become green.*

In fact, over the past two decades the lean community has focused on operational improvements to build a continuous improvement. In the lean model, work is based on the principles of continuous improvement, or Kaizen. Workers are responsible for identifying problems found on the production line and, in contrast to mass production, are able to stop the line for such problems. Floor workers are arranged in teams, with a team leader performing a coordinating role in addition to assembly tasks (Rothenberg 2001). A benefit of pollution prevention activities is that they are often value added for the firm since they reduce costs through material use reduction or through the avoidance of waste management costs (Florida 1996; Found 2009).

The next challenge for the lean community is to consciously account for the environmental issues. Gordon (2001) discusses some ways for integrating lean and green practices with a focus on cost reduction practices. The fundamental building block of lean thinking is continuous improvement, Kaizen, with its focus on problem solving and employee involvement fits with the notion of creating a greener industry. Therefore, the pursuit of continuous improvement, i.e. Kaizen, creates substantial opportunities for pollution prevention and waste and emissions reduction. Figure 4 illustrates the positioning of Lean and Green Manufacturing.

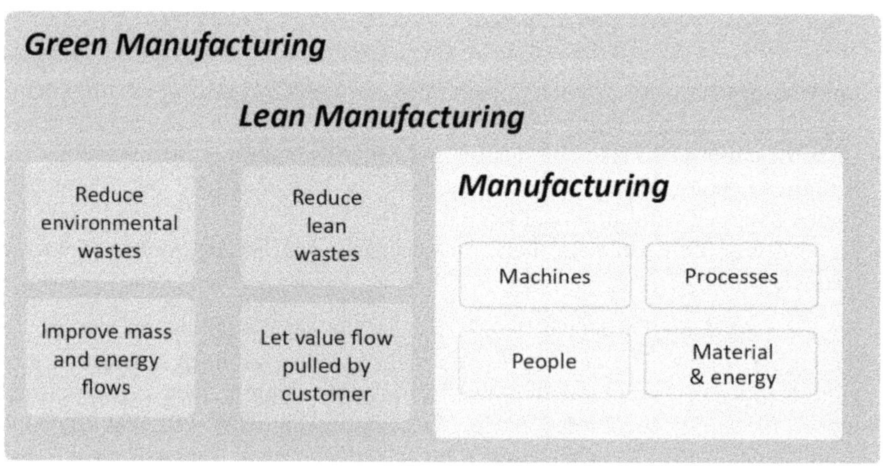

Fig. 4 Level of integration of lean and green with manufacturing processes. *Source* Developed by the authors

2.3.1 Integrating Lean and Green in Practice

Although the idea of lean thinking had not been fully explored by many in the environmental community, a number of articles were published where some of the lean thinking fundamentals, such as the need for people involvement (Venselaar 1995; Boyle 1999; Stone 2000; Remmen and Lorentzen 2000; Perron et al. 2006) the idea of learning by doing (Dieleman and Huisingh 2006) continuous improvement (Fresner 1998) and application of problem solving tools (Calia et al. 2009) were identified as necessary for implementing environmental policies by environmental researchers indicating a connection between lean and green practices.

To the environmental researchers that recognized the existence of lean thinking, there have been several initiatives discussing positive and negatives aspects of using lean to support the other dimension of sustainability, the environment, using different aspects and tools of lean for solving environmental problems and therefore contributing to a more sustainable business. For example, Chiarini (2014) identified that whilst Value Stream Mapping, TPM and other lean tools such as 5S and cellular manufacturing were positively associated with reducing environmental impact, there were no significant environmental savings as a result of Single Minute Exchange of Die (SMED). This raised an interesting debate on the typology of lean tools for different environmental impacts; similar to the Hines and Rich (1997) argument that there is a typology of Value Stream Mapping tools to identify production wastes.

Vais et al. (2006) in the study 'Lean and Green at a Romanian secondary tissue paper and board mill', present the development of technical environmental projects for accomplishing legal requirements and the use of lean tools, such as 5S's, Kaizen and autonomous maintenance for developing punctual improvements, optimizing the use of natural resources and production output.

EPA published The Lean and Environmental Toolkit in December 2006 (EPA 2006) to demonstrate that traditional lean tools can be applied to environmental wastes. This manual establishes guidelines for using existing lean tools for improving material flow to reduce the main flows that support the production process and that can affect the environment (such energy, chemicals, wastes, etc.).

Following these studies is possible to conclude that:

- *There are intrinsic linkages between lean and green—not least due to the relentless focus of lean on waste elimination.*
- *Lean tools and fundamentals are successful when used for promoting environmental improvements.*

According to Gustashaw and Hall (2008) an organization in which lean is already at the heart of its business system, and Kaizen the basis for continuous improvement culture, could adopt the same strategy for improving production energy and material flows. Deploying a lean strategy of improving the way that products and materials are sourced, manufactured, marketed and disposed of at the end of its life-cycle means that lean thinking can be used for creating a sustainable manufacturing. The authors state that by applying lean logic, for thermodynamic environmental improvement of mass-energy balances, the holistic improvement within a factory system boundary can benefit an existing business model greatly. Although this idea was stated by the Gustashaw and Hall (2008), no examples where found where pure lean thinking was expanded to create a new and integrated way of thinking. The examples found focus only on using and applying lean tools for promoting environmental improvement.

This idea of the existence of a new way of thinking, connecting the business thinking (such as lean) and the green dimension can also be sustained by analysing some pure green practices. Many researchers studied and proposed integrated approaches. Some of them explore the context of green manufacturing whilst others explore some limitation and success factors of cleaner production, pollution prevention initiatives expressing the need for a strategic approach, leadership support and integration into the existing business models. Table 3 presents a list of examples.

Building upon the existing work, some of the conclusions are:

- *Although lean and green integration was proven by many practical examples* (*Vais et al.* 2006; *EPA* 2006) *none of these have explored the idea of creating a new way of thinking.*

2.3.2 Dimensions of Lean and Green

Three components make up the 'triple bottom line' of corporate sustainability, a concept coined by Elkington (1998) which can be defined as meeting the needs of a firm's direct and indirect stakeholders (such as shareholders, employees, clients, pressure groups, communities etc.) without compromising its ability to meet the needs of future stakeholders as well (Dyllick and Hockerts 2002).

In 1999 Hawken, Lovins and Lovins discussed that there is a great potential of integrating lean thinking with its focus on QCD measures with environmental sustainability. Until recently lean manufacturing and the application of lean thinking has concentrated on the economic and some of the social aspects of sustainability. However, the essence of lean to produce more with less implies that lean thinking organizations use less resource, in the form of raw materials and energy. Therefore, lean thinking is green once it proposes the reduction of materials, wastes and energy that are required by the production; lean is creating a new manufacturing paradigm, which, inadvertently, includes an environmental sustainability element. Thus a Lean and Green Business Model makes this explicit and includes a deliberate, intended focus on reducing environmental impact that is measurable and forms part of a continuous improvement strategy.

According to Hall (2010) although lean thinking already explores some aspects of sustainability, people and profit, sustainability goes beyond this, including also the idea of environmental impact—mass and energy flows of everything that enters and leaves the system. Therefore, extending lean thinking to an integrated lean and green approach addresses the three core sustainability dimensions (people, profit and planet). A sustainable manufacturing business has to focus on eliminating wastes (profit) implementing Kaizen (people) and to explain the movement of mass

Table 3 Articles that explore the idea of integrated approaches to connecting business thinking (such as lean) and the green dimension

Key	Title	Authors	Year
Articles that propose the need of pollution prevention integration into existing systems	Cleaner production and profitability: analysis of 134 industrial pollution prevention (P2) project reports	Cagno et al.	2005
	Improving cleaner production by integration into the management of quality, environment and working conditions	Zwetsloot	1995
	New models of pollution prevention technical assistance	Atkinson	1994
Articles that explore the idea of integrated approaches to manufacturing—sustainable/green manufacturing	A study of the environmental management system implementation practices	Hui et al.	2001
	Modeling manufacturing evolution: thoughts on sustainable industrial development	Baldwin et al.	2005
	Material flows and environmental impacts of manufacturing systems via aggregated input–output models	Xue et al.	2007
	An integrated methodology for environmental impacts and costs evaluation in industrial processes	Santos da Silva et al.	2009
	A system model for green manufacturing	Deif	2011
Articles that explore some limitations and success factors of cleaner production, pollution prevention initiatives expressing the need for a strategic approach, leadership support and integration into the existing business models	Comparative evaluation of cleaner production working methods	van Berkel	1994
	The essential elements for successful cleaner production programmes	Zwetsloot et al.	1996
	To make zero emissions technologies and strategies become a reality, the lessons learned of cleaner production dissemination have to be known	Baas	2007
	Cleaner production: beyond projects	Baas	1995
	Limitations of cleaner production programmes as organizational change agents I. Achieving commitment and on-going improvement	Stone	2006

(continued)

Table 3 (continued)

Key	Title	Authors	Year
	Limitations of cleaner production programmes as organizational change agents. II. Leadership, support, communication, involvement and programme design	Stone	2006
	Strategic sustainable development—selection, design and synergies of applied tools	Robèrt et al.	2002
	Scenarios in selected tools for environmental systems analysis	Höjer et al.	2008

and energy within and through boundaries (planet) even if these boundaries are only a production cell, the entire factory or the whole supply chain.

- *A new way of thinking can be created by integrating pure lean thinking* (1)—*profit and* (2) *people with the dimension of green thinking* (3) *planet*

Therefore, based on the discussion of the five propositions stated earlier, this paper aims to propose a new and integrated way of thinking that (1) contributes and balances the three sustainability dimensions (people, profit and planet) and that (2) integrates to the pure lean thinking one new dimension, the environmental sustainability, the green thinking, developing a model that uses the Kaizen approach for dealing and improving environmental flows of mass and energy in manufacturing environment that already possesses a deployment level in applying lean.

2.4 Beyond Sustainability

And what comes after sustainability? Compression thinking (Hall 2010) may answer this question. With a top-level statement that establishes a sure survival of life and promotes quality of life using processes that work to perfection with self-correcting, self-learning systems, without the use of excess resources. With no wasted energy, no toxic releases and always quality over quantity, compression thinking is based on the fact that the society is near a turning point, the end of expansion. Population is expanding on an earth with finite resources. Traditional thinking from the industrial revolution and financial thinking need to be changed. So, the case for compression is based on 4 main drivers (1) Finite Resources, (2) Precarious Environment, (3) Overconsumption, (4) Pushback, as discussed by Hall (2010).

According to Hall (2010) lean thinking breaks a little from traditional thinking since its practitioners are used to removing waste from processes, not always represented by costs. But compression thinking has to step beyond this. Physical

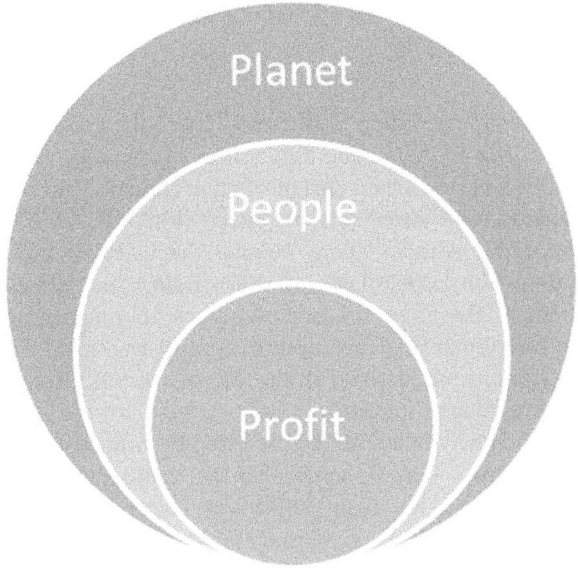

Fig. 5 The three vectors of sustainability viewed from compression thinking. *Source* Adapted from Hall (2011)

actions and their consequences must take priority over financial motivations. Therefore, compression begs for a fundamentally new economic thinking, looking beyond financial transactions to see the physical reality of what society and corporations do. Also, compression is not pure environmental. Environmental concerns are only one reason to make systemic changes. It calls for a different mind-set, for an integrated approach to deal with the increasing complexity of today´s work. Figure 5 shows how the three vectors presented by the sustainability concept are viewed based on compression thinking.

Following this, compression thinking brings a new way to see environmental issues. Differing from the sustainability concept, compression states that this should be part of bigger system, integrated into the core business model. Although compression has a much wider scope, it is understood that lean thinking is a way to get to compression.

3 The Lean and Green Business Model

In order to understand the purpose, the principles and the ways of working of the Lean and Green Business Model (L&GBM) and to explain why it is different than pure lean and pure green, the methodology section of this paper is divided four main blocks: The purpose of the model, (3.1); The principles of the model, (3.2); The ways of working of the model, (3.3) and why lean and green is different from pure lean or pure green thinking.

3.1 The Purpose of the Lean and Green Business Model (L&GBM)

Although, according to Bicheno (2000) the general purpose of lean thinking can be described in three main dimensions (1) Quality, (2) Delivery and (3) Cost, Lozano (2008), reviews the concept of environmental sustainability, established by several authors, and states that the green thinking can be quoted as the use of natural resources without going beyond the carrying capacities of the system and the production of pollutants without passing the biodegradation limits of the receiving system. Therefore, the general purpose of environmental thinking can be described in one dimension (Environment) with two main focuses: (1) Producing with the maximum productivity in the use of natural resources and (2) with the minimum environmental impact.

The idea of the L&GBM is using lean thinking to solve environmental problems, adding one more dimension to the traditional lean thinking, the Environment. In this context, the main objectives of the model are based on the fundamental building blocks of environmental sustainable practices:

- *Improving manufacturing processes resources productivity* by optimizing its supporting flows performance (materials and energy consumption and wastes generation);
- *Reduce manufacturing processes environmental impact*, by reducing all environmental wastes generated by production.

Following this, the L&GBM model can be defined as:

Producing exactly what the customer wants, exactly when (with no delay) at a fair price and with minimum waste and environmental impact by delivering the maximum productivity in the use of natural resources.

This means that lean and green thinking will be described in four dimensions, (1) Quality, (2) Delivery, (3) Cost) (4) Environment, linked to the three core sustainability dimension (1) Profit, (2) People and (3) Planet.

Figure 6 presents the position of L&GBM by locating it between pure lean and pure environmental thinking and illustrates how it integrates the sustainability vectors, in order drive towards compression. This begs fundamentally for a new economic thinking and calls for a different mind-set, for an integrated approach which deals with the increasing complexity of today's work.

3.2 The Principles of the L&GBM

In general terms, environmental thinking models, such as Industrial Ecology (Nielsen 2007; Tibbs 1992), Industrial Symbiosis (Boons et al. 2011), Eco-efficiency (Korhone 2007), Triple Bottom Line (Elkington 1997; Lenzen 2008), Natural

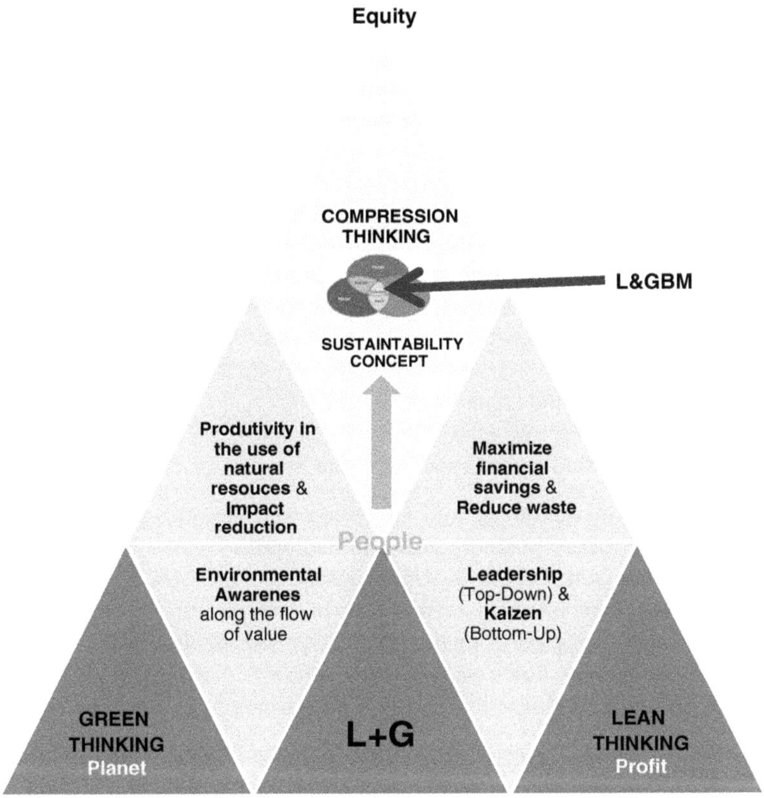

Fig. 6 Positioning the L&GBM

Capitalism (Robèrt 2002a; Hawken et al. 1999), The Natural Step (Robèrt 2002b) can be generalized by four common key principles:

- Identify environmental aspects and impacts;
- Measure environmental impact and the use of natural resources;
- Identify alternatives to reduce environmental impact and improve resource productivity;
- Continuous Improvement.

As stated previously, Womack and Jones (1996) offer five key principles for defining and describing lean thinking:

- *Specific value*: define value precisely from the perspective of the end customer in terms of the specific product with specific capabilities offered at a specific time;
- *Identify value streams*: identify the entire value stream for each product or product family and eliminate waste;
- *Make value flow*: make the remaining value creating steps flow;

- *Let the customer pull value*: design and provide what the customer wants only when the customer wants it;
- *Pursue perfection*: strive for perfection by continually removing successive layers of waste as they are uncovered.

To operationalize the lean thinking principles, the identification of value streams is key, to make value flow at the pull of the customer. In lean enterprises, manufacturing processes are organized in levels of flow, where:

- *The first level is the cell level*, the lowest production level in a manufacturing company organized by lean principals, composed by a finite number of operations/machines;
- *The second level is the factory level*, value stream level, composed by several cells that are part of the same value stream;
- *The third level is the extended value stream level*, composed by several sites (external supplier through to customer) that are part of the same value stream.

The leadership, methodology and execution patterns, designed for improving value stream performance in an organization that applies lean thinking as a strategy for increasing manufacturing performance, are used in the L&GBM. The difference here is that, instead of focusing in the flow of product (that is the main goal of improving manufacturing performance) the focus here is optimizing the use of the value stream supporting flows performance (mass and energy flows). Following this, the L&GBM can be described by five key principles:

- *Identify a stable value stream* (*VS*): Identify a stable value stream (level 1, 2 or 3). A stable value stream is a value stream that has improved and reduced the waste along the main dimensions of lean thinking (1) Quality, (2) Delivery and (3) Cost);
- *Identify the environmental impact* (*E*): Identify in the chosen value stream the environmental aspects and impacts;
- *Measure the environmental value streams* (*EVS*): Measure the value stream environmental impacts and the use of natural resources (the value stream supporting mass and energy flows);
- *Improve the environmental value streams* (*EVS*): Identify alternatives to (1) impact reduction and (2) resources productivity within the value stream;
- *Continuous Improvement* (*CI*): Set alternatives for improving the value stream supporting mass and energy flows.

Considering what was presented, Fig. 7 presents the overall idea of the L&GBM principles.

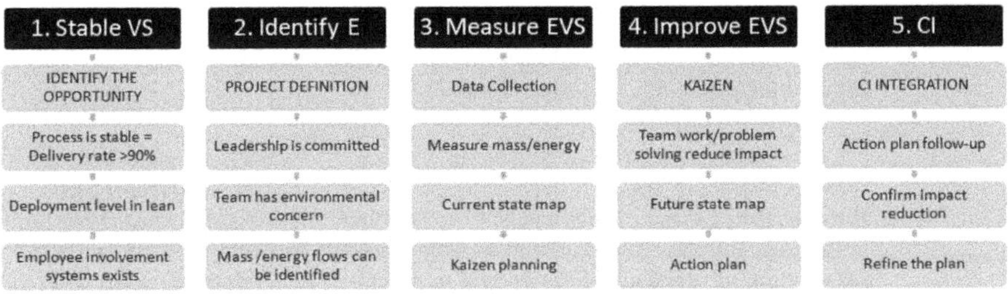

Fig. 7 The principles of the 5-step L&GBM. *Source* Developed by the authors

3.3 The Ways of Working of the L&GBM

The basic and most important idea of the L&GBM is that lean and green approaches will be integrated fully as part of the continuous improvement process of a manufacturing process, where the lean philosophy and ways of working were already in place, as described in the session 3.2. Following this, the objects of study of the L&GBM are the mass-energy flows of the manufacturing processes and the expected output for model application is the achievement of improvements in these thermodynamic flows (Materials, Chemicals, Water, Waste, Effluent, Energy) contributing to improvement of the overall performance.

One fundamental building block of lean thinking is continuous improvement, Kaizen, with its focus on problem solving and employee involvement, which fits perfectly with the notion of creating a greener industry. The L&GBM will be using the Kaizen approach for dealing and improving environmental flows of mass and energy of a manufacturing a cell and the value stream with the difference that here the focus is on optimizing the mass and energy flows.

3.4 Why the L&GBM Is Different Than Either Lean or Green Thinking?

Following the description of the L&GBM developed in sessions 3.1, 3.2 and 3.3, Table 4 highlights the fundamental differences of the L&GBM compared to pure green and pure lean thinking in terms of purpose, principles and ways of working of dealing of the sustainability vectors (People, Profit and Planet).

Considering what was presented in Table 4, L&GBM is different than pure *Green Thinking* due to:

- *L&GBM prioritizes the customer focus*: For being L&GBM it is necessary to be lean first; Therefore a prerequisite of deployment level in lean is key for Lean and Green;

Table 4 Table comparing Lean and Green with pure lean and pure green thinking

	Green environmental sustainability thinking	Lean lean thinking	L&GBM
General purpose	"Use of natural resources without going beyond the carrying capacities and the production of pollutants without passing the biodegradation limits of the receiving system" (Lozano 2008)	"Producing exactly what the customer wants, exactly when (with no delay), at fair price and minimum waste" (Bicheno 2000)	"Producing exactly what the customer wants, exactly when (with no delay), at fair price and minimum waste and environmental impact and the maximum productivity in the use of natural resources"
Main principals	1. Identify environmental aspects and impacts 2. Measure environmental impact and the use of natural resources 3. Identify alternatives to (1) impact reduction and (2) resources productivity 4. Continuous improvement	1. Specific value 2. Identify value streams 3. Make value flow 4. Let the customer pull value 5. Pursue perfection (Womack and Jones 1996)	1. Identify a stable value stream (level 1, 2 or 3); 2. Identify in the flow of value the environmental aspects and impacts 3. Measure value stream environmental impacts and the use of natural resources 4. Identify alternatives to (1) impact reduction and (2) resources productivity in value streams 5. Pursue perfection—continuous Improvement
People	1. Environmental awareness in all levels of the organization 2. High level of technical competence for people responsible for environmental impacts	1. Leadership (Top-Down) 2. Kaizen (Bottom-Up) People involvement and creation of solving problems culture	1. Leadership (Top-Down) 2. Kaizen (Bottom-Up) 3. Environmental awareness in along the flow of value
Profit	1. Equity (economic/environmental)	1. Maximize financial savings (revenue) 2. Reduce waste	1. Maximize financial savings (revenue) 2. Reduce waste (for all sources of wastes streams) 3. Equity (economic/environmental)
Planet	1. Productivity in the use of natural resources (mass and energy) 2. Environmental impact reduction (3R's)	None	1. Productivity in the use of natural resources (mass and energy) 2. Environmental impact reduction (3R's)

- *L&GBM identifies and measures environmental aspects and impacts based on value streams*: Traditional green thinking does not focus on the manufacturing ways of working to do this;
- *L&GBM focuses on a top-down and bottom-up approach*: for deploying environmental continuous improvements;
- *L&GBM prioritizes maximizing value and reducing costs*: It has an environmental approach, prioritizing financial savings and waste reduction as well.

Considering what was presented in Table 4, L&GBM is different to pure *Lean Thinking* due to:

- *L&GBM introduces into the traditional lean thinking a new dimension—the environmental concern aspect*: Traditional lean thinking focuses on three dimensions: Quality, Delivery and Cost. L&GBM introduces the environmental concern, requiring (i) minimization of the use of resources and (ii) reduction of environmental impact (iii) the need of environmental awareness along the flow of value;
- *L&GBM focuses on other sources of savings*: Traditional lean thinking considers only reduction of the seven classic wastes. With the introduction of the environmental variable concern along the flow of value, other sources of wastes may be focused and reduced, maximizing the overall savings.

The overall idea of the L&GBM encompasses the same principles of the lean thinking that are set in the house of lean, where the stability is the base, the Kaizen is it main pillar with the ultimate goal of improving performance, that in the case is based in three dimension, (1) Quality, (2) Delivery and (3) Cost. The difference here is that one more dimension, (E) Environment will be added to existing model. Figure 8 presents the idea.

Rich (2006) discusses lean improvement stages from chaos to control to competitive advantage, setting the natural steps to be followed by a manufacturing process implementing lean principles over a period of time. By concentrating first on stabilizing processes, where basic discipline, safety and morale is addressed and followed by improvements in quality, delivery performance and process flexibility, costs are reduced naturally, creating opportunities for further cost reduction that are realized in the later stages:

Process stability (Quality + Delivery + Flexibility) → Cost reduction

Perhaps, this logic does not take into consideration the other sources of cost that are part of the manufacturing process, the environmental wastes (materials and energy consumption and wastes generation) and that are not considered in the original Rich's model. Therefore, the L&GBM is built based on Rich's model, adding one extra variable to it:

Process stability (Quality + Delivery + Flexibility) + Environment → Cost reduction

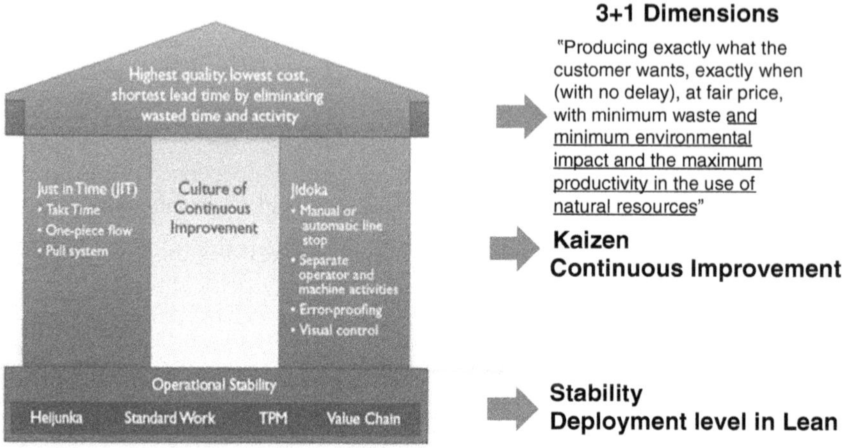

Fig. 8 The house of lean and L&GBM. *Source* Adapted from Rich (2006)

4 Results

The model was developed and implemented in a major, global engineering company that services the automotive and aerospace industries. Initially, the model was tested in two pilot cells in the automotive production plant in Brazil. These are termed Cell 1 and Cell 2 that represent two stages of the manufacturing process. Table 5 presents the basic characteristics of the manufacturing operational cells where the L&GBM was applied. The pilot Kaizen events were developed at the end of 2008 and 2010 and followed the five-step model. The Kaizen events each involved approximately 30 people, including all cell operators, leaders and managers, and maintenance people as well as environmental and lean specialists. The results of the Kaizen events are presented in Table 6. The main objectives of pilot testing were (1) confirm the 5 steps proposed and structure for the L&GBM for a cell before rolling it out for other several manufacturing cells of one manufacturing business, (2) confirm the prerequisites and participants, (3) analyze potential savings in terms of environmental improvements and cost reduction after applying the model, (4) identify model improvement opportunities, were all achieved. Therefore, the L&GBM for a cell was considered a good strategy for (1) improving manufacturing processes resources productivity by optimizing its supporting flows performance (materials and energy consumption and wastes generation) and for (2) reducing manufacturing processes environmental impact, by reducing all environmental wastes generated by production, the two main objectives of the L&GBM.

The action plans from both Kaizen workshops were not implemented totally because some of the ideas proposed by the Kaizen teams were considered not to be viable by technical experts following detailed technical analyses. However as

Table 5 Environmental and manufacturing characteristics of the pilot cells where the L&GBM was applied, including the application and evaluation of prerequisites

Manufacturing characteristics	Cell 1	Cell 2
Kaizen Date	November 08	Jun 10
Nature of operations	Steel machining	Assembly of manufactured parts
Main Cell Mass and energy flows	Energy Water Chemicals/oils Effluents Metallic waste Hazardous wastes	Energy Waste grease Hazardous wastes Cleaning cloths
Actual state data: energy and materials consumption and wastes generation	Energy consumption: 261 Mwh/month Water consumption : 1.4 m^3/month Chemicals usage : 0.6 m^3 /month Metallic wastes : 55 ton/month Hazardous wastes: 60 m^3/month	Energy consumption: 11 Mwh/month Waste grease: 0.2 ton/month Hazardous wastes: 3 m^3/month Cleaning cloths usage: 3120 units/month
Prerequisities		
Level of lean	Deployment level +	Deployment level +
Process stability	<90 % +	<90 % +
Application of employee involvement tools	In place +	In place +
Leadership support	High +	High +
Environmental awareness	In place +	In place +
Use of resources	High +	Medium -
Total cost of mass and energy flows (US$/Year)	1,005,000	483,500
Major impact in the Cell environmental cost	Metallic waste 68 %	Grease 75 %

Table 7 shows they were accepted substantially and generated significant benefits. Considering that the automotive company where the project is being tested has approximately 70 cells, if the model is implemented to the same extent in all cells, the L&GBM will generate a total annual savings of US$1,600,000/year for the company. Following the pilots, the L&GBM was then rolled out to other manufacturing cells. The rollout phase of the L&GBM was further developed in automotive manufacturing operations in Brazil in 2011. The model applied for this

Table 6 Results of Kaizen event: identification of improvement opportunities for the cell mass and energy flows

Cell impact	Cell 1	Cell 2
Energy saving (%)	8	6
General Chemical products consumption reduction (%)	91 (oils)	1 (grease in the product)
Water consumption reduction (%)	34	NA
Effluent generation reduction (%)	69	NA
Metallic waste generation reduction (%)	33	NA
Hazardous waste generation reduction (%)	67	45
Cleaning cloths Usage reduction (%)	NA	50
Grease waste generation reduction (%)	NA	100[a]
Average resources reduction (%)	50	40

NA not applicable

[a] 100 % waste elimination due to 100 % recycling of grease

Table 7 Implementation results

	Cell 1	Cell 2
% action plan implemented	94 %	81 %
Examples of improvement opportunity ideas that were identified during the Kaizen events that were implemented	*To reduce energy usage*: Motion sensitive and low energy lights were installed in low usage areas *To reduce metallic waste generation*: Forgings were redesigned for reducing machining and hence metallic waste *To reduce contaminated waste generation*: Plastic wrap containing oil contamination was eliminated from the containers—substantial savings in disposal costs	*To reduce energy usage*: All assembly cell lighting system was substituted to 54 W system that consumes less energy *To reduce grease waste generation*: (1) A new system was introduced to re-use the waste grease that was left in the used drums; (2) a new weighting standard was introduced in order to reduce the process waste grease
% Cost savings by reducing cell's mass and energy flows	13 % After implementing the action plans (1 year)	3 % After implementing the action plans (1 year)
Cost savings (US$/Year)	US$132,000 Results after implementing the action plans	US$15,000 Results after implementing the action plans

phase was unchanged from that applied previously. The Kaizen events followed the same structure as before. In total, through 2011, seven Kaizen events were held. Each of the seven manufacturing cells had different characteristics in terms of prerequisites for applying the L&GBM. The important point to highlight is that the pilot testing for the L&GBM was considered successful and it proved the business case for the L&GBM, confirming the proposed characteristics and prerequisites. The L&GBM was then rolled out to other manufacturing cells, including sisters' cells and for a value stream. Finally, it was applied outside of automotive and tested in plants in the US and UK.

5 Conclusions

Nature follows a distinct logic. From cradle to grave, the birth and death of every living thing, the composition of the atmosphere and the soil, the cycling of elements through air and waterways, and many other ecological assets are all the result of the evolution of living processes. The human species, while buffered against environmental immediacies by culture and technology, is ultimately fully dependent on the flow of ecosystem services and to the logic of the nature.

Manufacturing also has its own logic. In order to achieve competitive advantage, to be a lean enterprise, producing exactly what the customer wants, exactly when (with no delay) at fair price and minimum waste, the process of implementation follows a distinct logic: Cost reduction is ultimately fully dependent of stability, a function of quality, delivery and flexibility.

Lean and Green has its logic as well. In order not only to achieve competitive advantage but also to be a sustainable enterprise, producing exactly what the customer wants, exactly when (with no delay) at a fair price and with minimum waste and environmental impact by delivering the maximum productivity in the use of natural resources the process of implementation follows a distinct logic: To be lean and green it needs to first be a lean enterprise. It means that the lean and green thinking will be described in four dimensions, and not just three, (1) Quality, (2) Delivery, (3) Cost) (4) Environment, linked to the three core sustainability dimensions of (1) Profit, (2) People and (3) Planet.

L&GBM aims to (a) reduce environmental impact and (b) increase the productivity in the use of resources and observing the manufacturing structure, production flows and lean fundamental building blocks. Four key conclusions were identified while analysing the model. They are:

1. *L&GBM has a different purpose than traditional Lean or Environmental Thinking*: It introduces a new way of seeing environmental problems, (a) from the green side—prioritizing the customer focus, (b) From the lean side—including a new dimension in the traditional lean thinking—the environment.

2. *L&GBM covers the three dimensions of sustainability*: Since traditional lean thinking embodies two dimensions of sustainability (people and profit) L&GBM extends this and introduces another dimension to pure lean, the respect for the environment.
3. *L&GBM has a Lean to Green approach*: L&GBM application should be the continuation, a second step of a continuous improvement/lean culture already in place.
4. *L&GBM as an alternative approach to integrating environmental concern in operations management*: Since it uses lean fundamentals, L&GBM translates the environmental technical language to the manufacturing world.

In addition, the L&GBM shows that environmentally sustainable practices can be considered as an extension to a lean philosophy. Sustainability means meeting the needs of current generations without compromising the ability of future generations to meet their needs in turn. Three fundamental impacts, social, environmental, and financial (or People, Planet, Profit) evolved to define business objectives using the original Brundtland philosophy.

This concludes that lean leads us toward sustainability initiatives. Because it is much like lean in concept and practice, sustainability can be thought of as lean expanded to achieve a much broader objective. In a world of uncertainty about the economy and environment, where most corporate strategies do not contribute fully to the three pillars of sustainability, the L&GBM demonstrates the case for a new and innovative way of thinking for supporting the development and the evolution of a sustainable business.

References

Atkinson R (1994) New models of pollution prevention technical assistance. J Cleaner Prod 2 (2):101–106

Baas L (1995) Cleaner production: beyond projects. J Cleaner Prod 3(1–2):55–59

Baas L (2007) To make zero emissions technologies and strategies become a reality, the lessons learned of cleaner production dissemination have to be known. J Cleaner Prod 15(13–14):1205–1216

Baldwin J, Allen P, Winder B, Ridgway K (2005) Modeling manufacturing evolution: thoughts on sustainable industrial development. J Cleaner Prod 13(9):887–902

Beale J, Found P (2006) Human aspects of sustainable lean manufacturing: the relationship between employee attitudes, personality traits and behaviours. In: Proceedings of 16th international conference on flexible automation and intelligent manufacturing, 26–28 June 2006, vol 2. University of Limerick, Ireland, pp 1111–1117

Berger A (1997) Continuous improvement and Kaizen: standardization and organizational designs. Integr Manuf Syst 8(2):110–117

Berkel C (1994) Comparative evaluation of cleaner production working methods. J Cleaner Prod 2 (3–4):139–152

Bicheno J (2000) The lean toolbox. PICSIE Books, Buckingham, UK

Biggs C (2009) Exploration of the integration of Lean and environmental improvement. Ph.D. thesis, School of Applied Science, Cranfield University, UK

Boons F, Spekkink W, Mouzakitis Y (2011) The dynamics of industrial symbiosis: a proposal for a conceptual framework based upon a comprehensive literature review. J Cleaner Prod 19 (9–10):905–911

Boyle C (1999) Cleaner production workshops. J Cleaner Prod 7(1):83–87

Cagno E, Trucco P, Tardini L (2005) Cleaner production and profitability: analysis of 134 industrial pollution prevention (P2) project reports. J Cleaner Prod 13(6):593–605

Calia R, Guerrini F, Castro R (2009) The impact of six sigma in the performance of a pollution prevention program. J Cleaner Prod 17(15):1303–1310

Carleysmith SW, Dufton AM, Altria KD (2009) Implementing lean six sigma in pharmaceutical research and development: a review by practitioners. R&D Manage 39(1):95–106

Carson R (1965) Silent Spring. Penguin Books, London

Chiarini A (2014) Sustainable manufacturing-greening processes using specific lean production tools: an empirical observation from European motorcycle component manufacturers. J Cleaner Prod. doi:10.1016/j.jclepro.2014.07.080

Cobert C, Klassen R (2006) Extending the horizons: environmental excellence as key to improving operations. Manuf Serv Oper Manage 8(1):5–22

Deif A (2011) A system model for green manufacturing. J Cleaner Prod 19(14):1553–1559

Dieleman H, Huisingh D (2006) Games by which to learn and teach about sustainable development: exploring the relevance of games and experiential learning for sustainability. J Cleaner Prod 14(9–11):837–847

Dues C, Tan K, Lim M (2012) Green as the new lean: how to use lean practices as a catalyst to greening your supply chain. J Cleaner Prod 1:1–8

Dyllick T, Hockerts K (2002) Beyond the business case for corporate sustainability. Bus Strateg Environ 11(2):130–141

Elkington J (1997) Cannibals with Forks: the triple bottom line of 21st century business. Capstone, Oxford

Elkington J (1998) Cannibals with forks: the triple bottom line of 21st century business. Environ Qual Manage 8(1):37–51

EPA, United States Environmental Protection Agency (2006) The lean and environment toolkit. Available at: www.epa.gov/Lean. Accessed 17 Dec 2011

Florida R (1996) Lean and green: the move to environmentally conscious manufacturing. Calif Bus Rev 39(1):80–105

Found P, Beale J, Rich N (2006) A theoretical model for sustainable change. In: Proceedings of 16th international conference on flexible automation and intelligent manufacturing, 26–28 June 2006, vol 2. University of Limerick, Ireland, pp 1075–1082

Found P (2009) Lean and low environmental impact manufacturing. In: Proceedings of POMS 20th annual conference, Orlando, pp 126–130

Fresner J (1998) Starting continuous improvement with a cleaner production assessment in an Austrian textile mill. J Clean Prod 6(2):85–91

Gill A (2009) Estudo de Caso: fundamentação científica, subsídios para coleta de dados, como redigir o relatório. Atlas, São Paulo

Glavic P, Lukman R (2007) Review of sustainability terms and their definitions. J Cleaner Prod 15:1875–1885

Gordon P (2001) Lean and green: profit for your workplace and environment. Berrett-Koehler Publishers, San Francisco

Gustashaw D, Hall R (2008) From lean to green. AME Target Mag 24(5):6–14

Hall R (2010) Compression: meeting the challenges of sustainability through vigorous learning enterprises. CRC Press, Boca Raton

Hall R (2011) Compression thinking. In: Proceedings LERC annual conference 2011, Cardiff University, Lean Enterprise Research Centre, Cardiff, UK, pp 13–28

Hawken P, Lovins A, Lovins H (1999) Natural capitalism: creating the new industrial revolution. Little Brown & Company, Boston

Hines P, Rich N (1997) The seven value stream mapping tools. Int J Oper Prod Manage 17(1):46–64

Hines P, Holweg M, Rich N (2004) Learning to evolve: a review of contemporary lean thinking. Int J Oper Prod Manage 24(10):994–1011

Hui J, Chan A, Pun K (2001) A Study of the Environmental management System Implementation Practices. J Cleaner Prod 9(3):269–276

Höjer M, Ahlroth S, Dreborg K-H, Ekvall T, Finnveden G, Hjelm o, Hochschorner E, Nilsson M, Palm V (2008) Scenarios in selected tools for environmental systems analysis. J Cleaner Prod 16(18):1958–1970

Hopp W, Sperarman M (2008) Factory physics. McGraw-Hill, Irwin

King AA, Lenox MJ (2001) Lean and green? An empirical examination of the relationship between lean production and environmental performance. Prod Oper Manage 10(3):244–256

Korhone J (2007) From material flow analysis to material flow management: strategic sustainability management on a principle level. J Cleaner Prod 15(17):1585–1595

Lander E, Liker J (2007) The Toyota production system and art: making highly customized and creative products the Toyota way. Int J Prod Res 3:2681–2698

Lenzen M (2008) Sustainable island businesses: a case study of Norfolk Island. J Cleaner Prod 16 (18):2018–2035

Liker JK (2004) The Toyota way: 14 management principles from the world's greatest manufacturer. McGraw-Hill, New York

Liker JK, Meier DP (2007) Toyota talent: developing your people the Toyota way. McGraw-Hill, New York

Liker JK, Hoseus M (2008) Toyota culture: the heart and soul of the Toyota way. McGraw-Hill, New York

Liker JK, Convis GL (2012) The Toyota way to lean leadership: achieving and sustaining excellence through leadership development. McGraw-Hill, New York

Lovelock JE (1990) Hands up for the Gaia hypothesis. Nature 344(6262):100–102

Lozano R (2008) Envisioning sustainability three-dimensionally. J Cleaner Prod 16:1838–1846

Lozano R (2012) Towards a better understanding of sustainability into companies' systems: an analysis of voluntary corporate initiatives. J Cleaner Prod 25:14–26

Mann D (2005) Creating a lean culture: tools to sustain lean conversions. Productivity Press, New York

Maxwell J, Rothenberg S, Schenck B (1993) Does lean mean green? The implications of lean production for the environmental management. MIT working paper

Meadows DH, Meadows DL, Randers J, Behrens W (1972) The limits to growth. Universe Books, New York

Nielsen S (2007) What has modern ecosystem theory to offer to cleaner production, industrial ecology and society? The views of an ecologist. J Cleaner Prod 15(17):1639–1653

Ohno T (1988) Toyota production system: beyond large-scale production. Productivity Press, Oregon

Osborn AF (1963) Applied imagination: Principles and procedures of creative problem solving. Charles Scribner's Sons, New York

Perron G, Côté P, Duffy F (2006) Improving environmental awareness training in business. J Cleaner Prod 14(6–7):551–562

Porter M, Van Der Linde C (1995) Toward a new conception of environmental competitive relationship. J Econ Perspect 9(4):97–118

Remmen A, Lorentzen B (2000) Employee participation and cleaner technology: learning processes in environmental teams. J Cleaner Prod 8(5):365–373

Rich N (2006) Understanding the lean journey. In: Rich N, Bateman N, Esain A, Massey L, Samuel D (eds) Evolution of lean: lessons from the workplace. Cambridge University Press, Cambridge, pp 11–31

Robèrt K-H (2002a) The natural step story. Seeding a quiet revolution. New Society Publisher, Gabriola Island, BC

Robèrt K-H (2002b) Strategic sustainable development: selection, design and synergies of applied tools. J Cleaner Prod 10(3):197–214

Rothenberg S (2001) Lean, green and the quest for superior environmental performance. Prod Oper Manage 10(3):228–243
Rother M, Shook J (2003) Learning to see. Lean Enterprise Institute, Cambridge, MA
Silva P, Amaral F (2009) An integrated methodology for environmental impacts and costs evaluation in industrial processes. J Cleaner Prod 17(15):1339–1350
Sobek II DK, Smalley A (2008) Understanding A3 thinking: A critical component of Toyota's PDCA management system. CRC Poductivity Press, Boca Raton, FL
Sugimori Y, Kusunoki K, Cho F, Uchikawa S (1977) Toyota production system and Kanban system: materialization of just-in-time and respect-for-human system. Int J Prod Res 15 (6):553–564
Stern N (2007) The economics of climate change: the stern review. Cambridge University Press, Cambridge
Stone L (2000) When case studies are not enough: the influence of corporate culture and employee attitudes on the success of cleaner production initiatives. J Cleaner Prod 8(5):353–359
Stone L (2006a) Limitations of cleaner production programmes as organizational change agents. I. Achieving commitment and on-going improvement. J Cleaner Prod 14(1):1–14
Stone L (2006b) Limitations of cleaner production programmes as organizational change agents. II. Leadership, support, communication, involvement and programme design. J Cleaner Prod 14(1):15–30
Tibbs H (1992) Industrial ecology: an environmental agenda for industry. Whole Earth Rev Winter 4–19
Vais A, Miron V, Pedersen M, Folke J (2006) Lean and Green at a Romanian secondary tissue paper and board mill—putting theory into practice. Resour Conserv Recycl 46:44–74
Venselaar J (1995) Environmental training: industrial needs. J Cleaner Prod 3(1–2):9–12
Womack J, Jones D (1996) Lean thinking. Free Press, New York
World Commission on Environment and Development (OCED) (1987) Our common future. Available at: http://www.un-documents.net/wced-ocf.htm. Accessed on 17 Dec 2011
Xue H, Kumar V, Sutherland J (2007) Material flows and environmental impacts of manufacturing systems via aggregated input–output models. J Cleaner Prod 15(13–14):1349–1358
Yang M, Hong P, Modi S (2011) Impact of lean manufacturing and environmental management on business performance: an empirical study of manufacturing firms. Int J Prod Econ 129:251–261
Zokaei K, Martinez, F, Vazquez D, Evans B, Sarmadi K, Peattie K, Pampanelli A, Hines P, Trivedi N (2010) Best practice tools and techniques for carbon reduction and climate change: lean and green report. Lean Enterprise Research Centre, Cardiff University, Cardiff
Zwetsloot G (1995) Improving cleaner production by integration into the management of quality, environment and working conditions. J Cleaner Prod 3(1–2):61–66
Zwetsloot G, Geyer A (1996) The essential elements for successful cleaner production programmes. J Cleaner Prod 4(1):29–39

Sustainability Through Eco-Design: Shedding Light on the Adoption of the ISO 14006 Standard

Beñat Landeta Manzano, Germán Arana-Landín, Patxi Ruiz de Arbulo and Pablo Diaz de Basurto

Abstract The ISO 14006 ecodesign standard was created in 2011 with a view to complementing the successful ISO 14001 standard for company environmental management. Although there are a lot of works in literature on the subject that have analysed the adoption process for this last-mentioned standard, the former has not actually received so much attention. To try and make a contribution to special literature, the aim of this article is to carry out an exploratory analysis of how the ISO 14006 standard has been adopted and integrated within a series of Spanish companies that have been pioneers in doing so. After analysing the main motives that have led companies to adopt the ISO 14006 standard, the study then attempts to shed light on the impact of such adoption both on an operative level and in terms of company results. A series of conclusions are also gathered in this chapter that may be of interest both to companies and to other stakeholders.

Keywords Ecodesign · ISO 14006 · Empirical study · Spain

B. Landeta Manzano · P. Ruiz de Arbulo · P. Diaz de Basurto
Faculty of Engineering, University of the Basque Country UPV-EHU,
Alameda Urquijo S/N, 48013 Bilbao, Spain
e-mail: benat.landeta@ehu.es

P. Ruiz de Arbulo
e-mail: patxi.ruizdearbulo@ehu.es

P. Diaz de Basurto
e-mail: pablo.diazdebasurto@ehu.es

G. Arana-Landín (✉)
Politechnical School College, University of the Basque Country UPV-EHU,
Plaza Europa 1, 20018 San Sebastian, Spain
e-mail: g.arana@ehu.es

A. Chiarini (ed.), *Sustainable Operations Management*,
Measuring Operations Performance, DOI 10.1007/978-3-319-14002-5_8

1 Introduction

The adoption of Environmental Management Systems (EMSs), notably the ISO 14001 standard (ISO 2004), has grown significantly worldwide over the last two decades (for a recent review, see Heras-Saizarbitoria and Boiral 2013). Following this successful path, some other management standards that deal with environmental management aspects have been launched, the ISO 14064 standard for quantification and reporting of greenhouse gases (ISO 2006a, b, c; Stechemesser and Guenther 2012), the ISO 14031 standard for environmental performance evaluation (ISO 2013a, b; Comoglio and Botta 2012), the ISO 50001 standard for energy management (ISO 2011b; McKane 2010), among many others, and the ISO 14006 standard for ecodesign (ISO 2011a; Arana-Landin and Heras-Saizarbitoria 2011; Rio et al. 2013), between others.

ISO 14006 was created with a view to complementing the ISO 14001 standard (Cluzel 2012; Savita et al. 2012). It establishes certain requirements governing the working system that need to be met by the company with regard to activities that have an environmental impact (ISO 2011a). The main feature of the ISO 14006 ecodesign standard involves the compiling of a series of guidelines regarding incorporating ecodesign into an EMS, which enables the environmental variable to be systematically incorporated into the design and development process of products and/or services (ISO 2011a). Integration of Life Cycle Thinking in company management is a basic objective of the standard, i.e. integrating the analysis and assessment of environmental impact caused by products throughout their life cycle into the design process (Sanyé-Mengual et al. 2014; Salomone et al. 2013).

Taking into consideration these aspects, following this introduction, the article goes on to a literature review of research related with the adoption process of ecodesign standards. It later continues in the third section to describe the empirical study, while in the fourth section, the results are shown. The fifth section contains the discussion and conclusions of interest to those groups involved in the adoption of this type of standard, and the references are provided in the sixth- and last- section.

2 Literature Review

The ISO 14006 standard would not appear to enjoy the popularity of other environmental management standards (ISO 14001, EMAS or ISO/IEC 27001), due to the fact that it is a young, very specific standard governing management of a process—only of interest to those organisations that carry out design activities (Landeta et al. 2013). Furthermore, ISO 14006 "…does not establish by itself specific environmental performance criteria, and is not intended for certification purposes" (ISO 2011a).

This has had an effect on academic literature among which numerous studies can be found regarding the use of specific ecodesign tools and adoption of environmental management (Sanyé-Mengual et al. 2014; Bovea and Pérez-Belis 2012).

However, there are far fewer references regarding the adoption of ecodesign management standards by companies in comparison to adoption of ecodesign techniques (Pigosso and Sousa 2011; Knight and Jenkins 2009), and even fewer that analyse the implications that this entails for companies and the impact on results (Deutz et al. 2013; Kara et al. 2005; Arana-Landin et al. 2013). Specifically, with regard to the ISO 14006, available references are extremely rare. Among the ones existing, attention should be drawn to studies by Arana-Landin and Heras-Saizarbitoria (2011), Arana-Landín et al. (2012) and Landeta et al. (2013), which, using case studies, have noted that companies would seem to be satisfied with having adopted the standard UNE 150301 (the standard used as a reference point for creating the ISO 14006). This point was shared by previous studies related to ecodesign tools that had been carried out in manufacturing (Collado-Ruiz et al. 2008; Justel-Lozano 2008) and building (Chen et al. 2010) companies. In particular, companies maintain they have managed to reduce the environmental impact of their products, mainly improving the energy efficiency (Muñoz et al. 2009; Arana-Landin and Heras-Saizarbitoria 2011; Hischier and Baudin 2010; Sanchez et al. 2007), although many aspects that might be improved were not able to be dealt with owing to the increase in costs that this entailed. Furthermore, in a unique case study Fernández (2012) point out the company obtained operational improvements after adopting the standard. Lastly, Knight and Jenkins (2009), for the manufacturing sector, and Landeta et al. (2013) and Arana-Landin et al. (2012), for the building sector, state that companies have obtained environmental improvements despite the fact the customer rejected some of the improvements put forward, as this meant an increase in the initial cost. Even though the price increase was easily offset if we take into account the savings gained during the usage phase. This aspect was pointed out by Luttropp and Lagersted (2006). They emphasized a sustainable and environmental protection perspective may even have negative effects, if the price does not vary.

Taking into account these previous studies, the aim of this research has been to carry out an exploratory analysis, based on the motives behind companies doing so, of how the ISO 14006 standard has been integrated within their own management standard, and its operative impact and results.

3 Empirical Research

In order to address the questions raised, an empirical study of a qualitative nature was designed, based on case studies. This methodology was selected because of its suitability when analyzing the complex process of ISO 14001 adoption, in which—as has already been stated—diverse agents and actors interact (Heras-Saizarbitoria and Boiral 2013).

Research of a descriptive, mainly exploratory, nature was planned, to facilitate greater penetration in and understanding of the subject, so as identify propositions that are capable of generalization in relation to the practices observed (Eisenhardt 1989;

Yin 2009). The fieldwork was carried out in Spain, and was developed between February 2012 and March 2014, and had two main components. First, a series of semi-structured in-depth interviews were conducted with the managers in charge of ISO 14006 in the organization. The interviews revolved around a semi-structured script. This approach proved to be sufficiently open, and was consistent with the inductive method for analyzing information chosen and, consequently, did not distort the evidence obtained. Second, the organizations analyzed made a very broad range of documentation available for research related to the management system and that in this article we'll present just very shortly.

A decision was made to study the cases of companies belonging to the four industrial sectors most present among those certified in ecodesign in the Basque Country as of December 2011, i.e. furniture manufacturers and firms from the chemical, electrical-electronic and capital equipment sectors. 13 cases were developed in total: 3 according to industrial sector, except for those corresponding to electrical-electronic equipment in which case 4 were developed (Table 1).

Those cases of companies were chosen that have the most proven track record in ecodesign projects and those that offered the most accessibility to information. An attempt was also made to maintain a balance between the number of cases analysed and the capacity for treatment of data and management of information by the researcher, thus ensuring the information would be sufficient for the purpose of carrying out a cross-sectional analysis of the cases.

As recommended in specialist literature on the subject (e.g. Yin 2009), the validity of factors was supported in the course of the research by the use of diverse sources of information (direct observation, consultation, interviews, and internal and external documentary information). Internal validity was guaranteed by the search for common patterns that explain the phenomena, while reliability was ensured by using semi-structured interviews of the same type and with an assessment protocol of cases against each factor.

4 Results

Below is provided a comparative analysis of the behaviour of the companies subject to study from the four industrial sectors (chemical, electrical-electronic, consumer goods and furniture). The cross-sectional comparative analysis carried out in accordance with the comparison and contrasting techniques proposed by Miles and Huberman (1994) enabled the extent of compliance with the research proposals to be ascertained and some conclusions to be drawn for the set of industrial sectors.

Table 2 shows a summary of the results obtained from the case study for each of the sectors subject to study in terms of the research proposals. In the first column, from the left of the table, are listed the research proposals, classified into three groups: motives, the adoption and certification process, and the results obtained. The other columns refer to key aspects selected from each of the groups according to the work proposals put forward.

Table 1 Characteristics of interviewed companies

Company	Size	Type of product	Sector	Certificates in accordance with management standards
Akaba	SME (small)	Office and commercial furniture	Furniture	ISO 9001, "EKOSCAN", ISO 14001, ISO 14006
Ofita interiores	SME (medium)	Office and commercial furniture	Furniture	ISO 9001, PECAL 9, ISO 14001, OHSAS 18001, ISO 14006
Euro seating international	SME (medium)	Seats for communities	Furniture	ISO 9001, ISO 14001, ISO 14006
Inteman	SME (small)	Chemical and biological products for general industry	Chemical	ISO 9001, ISO 14001, EMAS, ISO 14006
A&B laboratorios	SME (small)	Chemical and biological products for general industry	Chemical	ISO 9001, ISO 14001, UNE 166002, ISO 27001, ISO 14006
DTS-OABE	SME (micro)	Biocides	Chemical	ISO 9001, "EKOSCAN", ISO 14006
Soraluce	SME (medium)	Boring and milling machinery	Capital equipment	ISO 9001, EFQM (SILVER), ISO 14001, OHSAS 18001, ISO 14006
Gamesa	Large enterprise	Wind turbines	Capital equipment	ISO 9001, ISO 14001, ISO 50001, ISO 14006
Orona	Large enterprise	Accessibility and mobility solutions (elevators, escalators…)	Capital equipment	ISO 9001, ISO 14001, ISO 14006
BSH electrodomésticos	Large enterprise	Household electrical appliances	Electrical-electronic	ISO 9001,ISO 14001, EMAS, UNE 166002, ISO 14006
Fagor electrodomésticos	Large enterprise	Household electrical appliances	Electrical-electronic	ISO 9001, EFQM (GOLD), ISO 14001, EMAS, OHSAS 18001, ISO 14006
Geyser-Gastech	SME (medium)	Domestic small water-heaters	Electrical-electronic	ISO 9001, ISO 14001, OHSAS 18001, ISO 14006
ABB Niessen	Large enterprise	Electrical equipment for industrial and household use	Electrical-electronic	ISO 9001, ISO 14001, ISO 14006

Source Put together by the authors

Table 2 Summary of results obtained from the case study of the furniture, chemical, consumer goods, and electrical-electronic sectors in terms of the work proposals put forward

	Proposal	Furniture	Chemical	Capital equipment	Electrical-electronic
Motives					
1	Customers	Meet the needs and requirements of environmentally-friendly products on especially informed international markets	Growing environmental awareness-raising of customers	Growing interest on the part of the market in more energy-efficient products	Increasingly demanding market that is concerned with the environmental behaviour of products
2	Product/service image	Obtain a product that can reinforce the company's brand strategy	Reinforce brand credibility	Necessary to reinforce the image of quality, efficiency and exemplary environmental behaviour on the part of the product	Improve product image
3	Company image	Differentiating factor within a highly competitive environment	Become detached from the image of a polluting and dangerous sector	There is a widespread and permanent need for differentiation in terms of competitors	Improve company image
4	Access new markets	Reinforce exports with products that have high added value in which the decision to acquire them is not so decisive	New market niches for biotechnological products on the European market	Approach to sectors which make high environmental demands of suppliers	Help to become consolidated and even access new markets on which energy efficiency and the cost of residue management are becoming competitive advantage factors
5	Action taken by public administrative bodies	Promotion of ecodesign on a strategic level by environmental and innovation agencies attached to regional governments	The drive and technical support offered by Ihobe have been keys	The drive and technical support offered by Ihobe have been keys	Attempting to reinforce legislative monitoring in environmental matters and anticipating future environmental legislation have been keys

(continued)

Table 2 (continued)

	Proposal	Furniture	Chemical	Capital equipment	Electrical-electronic
6	Improving the environmental impact of products	Company policy based on continuous improvement and excellence, not just regarding processes but also the product throughout its life cycle (life cycle thinking)	Continuous improvement of chemical and biological solutions to ensure they are cleaner and safer	Ecodesign might be a tool to help foster the development of eco-efficient products	Need to develop new systems that are more efficient and innovative
7	Environmental awareness-raising	Conservation of the environment is a strategic commitment	Sustainable development model for the business which entails developing effective and safe products for the user and the environment	Committed to the development a sustainable business model	Major conviction regarding a business model that takes into consideration all aspects of sustainability
Process					
8	Previous experience in processes involving adoption and certification of management systems	Adoption and certification tradition in accordance with international standards governing quality assurance, the environment and safety at work	Adoption and certification tradition in accordance with international standards governing quality assurance, the environment and safety at work	Previous experience in adoption and certification in accordance with international standards governing quality assurance, the environment and safety at work has been a key element in facilitating the process	Previous experience in adoption and certification in accordance with international standards governing quality assurance, the environment and safety at work has been a key element in facilitating the process
9	Difficulties with identification and assessment of environmental aspects and impact	Difficulties with obtaining initial data from suppliers of materials and components in order to assess environmental impact. Main motives: lack of awareness about the concept and implications of ecodesign, distrust of and difficulties with obtaining the	Difficulties with obtaining initial data from suppliers of substances in order to assess environmental impact. Main motives: lack of awareness about the concept and implications of ecodesign, distrust of and difficulties with	The large amount and diversity of associated materials, processes and components has meant a considerable effort required to put together an inventory; difficulties with using LCA commercial tools and their integration with other	Incomplete commercial data bases and unsuitable LCA commercial tools deemed unsuitable for company needs

(continued)

Table 2 (continued)

	Proposal	Furniture	Chemical	Capital equipment	Electrical-electronic
		environmental information required	obtaining the environmental information required	existing systems in the company	
Results					
10	Improvement in financial results	The influence of ecodesign adoption in operative costs has not been quantified; widespread belief in an increase in costs during the design and development phase	Major acceptance of the eco-designed product on the mainly European market; the eco-designed product is, generally speaking, more expensive than a similar conventional product, to a large extent owing to the costs of raw materials and the greater effort required in the product's design and development process	Ecodesign has helped the company to become consolidated on the market and, in the case of Orona, has even resulted in products being obtained that have given rise to significant success in terms of sales	Slight reduction in operative costs, mainly owing to savings in consumption of materials and shipment
11	Increase in added value	Contributes towards continuous improvement of processes and products with the inclusion of the environmental variable	Constitutes one more innovation tool; it entails taking other parameters into consideration in design and development which would have previously not have been the case; the end product offers greater performance features	Is a suitable innovation tool; it fosters improvements in key factors attached to the product such as functional nature, reliability or durability, energy efficiency and a reduction in residue generated	Provides another perspective to product creation, and fosters innovation in the product

(continued)

Table 2 (continued)

	Proposal	Furniture	Chemical	Capital equipment	Electrical-electronic
12	Improvement in their product/service image	Improves product image, although international markets tend towards product certification	Improves product image and makes it easier to obtain ecolabels	Certification reinforces brand image	Reinforces brand image
13	Improvement in their company image	Improves the company image	Lends support to the message put out by the manufacturer and provides it with credibility in the eyes of the customer; has enabled the company image to be improved and to differ from the competition	The certificate is a differentiating factor within the market	Improves the company image and even enables it to differ from the competition
14	Anticipation of compliance with legislation	This is a tool that makes it easier to anticipate future environmental legislation	Has fostered compliance with future environmental legislation	Has entailed anticipation of compliance with future legal demands both on a national and international level	Has fostered anticipation of compliance with future environmental legislation
15	Satisfaction with results	They seem satisfied, except for Orona, which would not opt for certification; preference for product certification, because the standard does not establish a system for recognising environmental behaviour of the product based on informing the customer	They seem satisfied; provides another perspective to product development (life cycle thinking); there is no clear commitment towards demanding that a level of specific requirement be established for compliance with ecodesign guidelines in future reviews of the ISO 140067	They seem satisfied; companies think they have helped mainly to make improvements in the creation process and the result is a product with greater added value; however, companies demand that a level of specific requirement be established for compliance with ecodesign guidelines in future reviews of the ISO 140067	They seem satisfied; it is a competitive tool for fostering continuous improvement in the environmental behavior of products

Source Put together by the authors

However, discussion of results provided below adds data which is not shown in the table which, although not referring to the work proposals, does nonetheless refer to other aspects attached to motives, the process and the results obtained from adoption and certification. In this respect, we should also recall for clarification purposes that the case study method enables a wealth of data to be captured in detail about the problem with research, and provides researchers with the flexibility they need to explore additional issues raised by those interviewed (Soosay et al. 2008).

4.1 Motives

In the chemical sector, obtaining more sustainable products for three of the companies analysed is dependent on obtaining products that ensure maximum safety at work, as environmental aspects in this sector are closely related to safety, e.g. both aspects are improved by replacing chemical substances with biological ones.

In the electrical-electronic and consumer goods sectors for instance, the launching on the market of more sustainable environmental products is linked to a concern with creating products which are more energy efficient, insofar as this is actually an aspect that the customer rates highly owing to its impact on the costs of using the product during its life cycle. Yet for Fagor and ABB Niessen the main motive was the need to adapt more quickly to increasingly more demanding and complex environmental legislation within an environment marked by the globalisation of markets.

Other motives such as a reduction in operative costs or attempts to better meet and improve customer expectations also formed part of business culture. However, adopting a series of strategic commitments in all industrial firms aimed at creating a new, more sustainable and environmentally-friendly business model of development, and improving environmental performance not only of the company but also of the products they market, were generally speaking noteworthy aspects cited.

4.2 Adoption and Certification Process

The adoption and certification process for the ISO 14006 standard was relatively simple in all cases. All industrial firms had experience in adopting management systems (ISO 9001, ISO 14001, EMAS, OHSAS 18001, etc.) which had created a working culture based on defined, orderly and systematised processes, and integrated ecodesign into the organisation's management system. This to a large extent ironed out any possible reticence to adoption of the ecodesign standard among staff —"this is the customary reaction to the introduction of new management systems," they point out at Geyser Gastech. In any event, all the companies highlight the effort made to convince and motivate everyone in the organisation to gain their support, which proved essential for the success of the process, although in the chemical sector, the sales network proved to be especially reluctant—specifically in the larger organisations such as Inteman and A&B Laboratorios. They did not see the value of ecodesign as a sales argument. However, ecodesign was linked to giving up using certain chemical products, thus increasing safety, and this aspect started to attract the sales network.

In this respect, management involvement in all companies proved very important, albeit not on the same level in all cases. Attention should be drawn to the greater involvement in especially innovative companies—those from the chemical and furniture sectors.

The management was responsible for providing tools and people to ensure the project would be successful, appropriately choosing those in charge of the standard adoption process and providing any external assistance deemed necessary with contrasting experience for preparation and advice throughout the process. All companies point out that such advice had been the key throughout the process. At Gamesa they also added that they had missed the support of certifying bodies less and stated that the experience gained by certification of the UNE 150301 (AENOR 2003) ecodesign standard remained insufficient. Moreover, the fact that the standard had not been created for certification purposes raised doubts, especially among companies, but also among the different interest groups.

Another major difficulty that companies have had to overcome and are still working on—especially in the electrical-electronic and consumer goods sectors—has been the lack of staff training, in particular in the most involved areas such as design and development. There were doubts and uncertainty that made it difficult to adopt environmental criteria during the design and development phase and, in the opinion of companies belonging to the consumer goods sector, the job market has not responded to the demand for professionals with a command of ecodesign.

Companies highlight the discovery of Life Cycle Assessment (LCA) as a methodology for assessing the environmental impact of the product throughout its life cycle, which has proved a very useful discovery for design and development technicians. It enables them to improve the product's integral perspective and ascertain the environmental influence of their decisions throughout its entire life cycle.

As regards this analysis, the main problems refer to calculation of the product's environmental impact. One of them is the absence of initial data to enable the impact of the product of reference to be assessed, especially in the first experience in ecodesign. Both then as now, all companies would experience serious difficulties with obtaining the information required of its suppliers in order to put ecodesign into practice. In this sense, consensus exists when pointing out that it is genuinely difficult to achieve traceability of data regarding the environmental aspects of materials and components (inputs).

Normally these are small firms with a limited capacity for attending to the information needs required of them with regard to design. Furthermore, at this time of the current serious crisis, they generally have fewer resources at their disposal to enable them to do so, and with suppliers confining themselves to complying with the legislation applicable to them. The smaller companies subject to study, such as

Akaba and DTS OABE, add that the relatively small volume of purchases from certain suppliers implies that they are unable to exert a major influence over them. This problem would seem to be less important in large companies such as Gamesa, BSH and Fagor. In such cases, the influence they exert on suppliers in demanding a greater environmental commitment proves to be a determining factor—even more so if this is a widespread trend on the market.

Companies from the chemical sector have been especially active in developing data bases and LCA tools, except for DTS OAB, which is perhaps restricted owing to availability of resources. This is a sector in which over 100,000 chemical substances are marketed in Europe and commercial data bases only have a really small proportion of substances with environmental data at their disposal.

The second problem linked to the calculation of environmental impact is the absence of a flexible and truthful tool for doing so. Apart from Inteman and A&B Laboratorios, both chemical firms, only Orona from the consumer goods sector has developed LCA software specifically for environmental indicators. This has enabled its ecodesign tools and the ecodesign process itself to be better integrated into activities associated with the area of product design and development.

The other companies have opted to use data bases and commercial LCA tools, although have experienced integration problems with the tools used by the area of product design and development. In some cases, it is shown that ecodesign is not fully integrated into company activities, as in the case of Fagor, where it was applied in certain projects but in practice has not been internalised in the working system within the area of design. Most companies have changed LCA software on more than one occasion so as to ensure greater integration into their design system and adaptability to their requirements.

Moreover, all companies stress that the ISO 14006 standard determines how environmental improvements can be quantified—each manufacturer can choose the comparative reference and methodology and calculation tools as they see fit, in accordance with their own criterion. The criterion used to choose the basis for comparison is an open one, as the environmental balance may prove to be positive or negative depending on the model chosen. Neither does it establish mechanisms for controlling the effectiveness of the ecodesign methodology for products applied by the company, which implies that the overall environmental impact of the product will vary. As no common comparative reference exists, companies face a problem when trying to highlight ecodesign on the market. The chemical firms Inteman and A&B Laboratorios—two of the most active in terms of the continuous development of their external communication methods—created their own label to identify the eco-designed product and a classification system for the product's environmental impact, very similar to the energy label used in the case of Inteman for taking advantage of the fact that customers had been familiarised with it and found it easy to understand. Likewise, this is the main reason that would seem to encourage companies from the chemical and furniture sectors to opt for environmentally-friendly labels for their products. They state that this is the trend on international markets and represents a form of recognition for those products.

Another limitation of the standard refers to the margin for environmental improvement in the re-design of eco-designed products. Companies are facing the challenge of updating eco-designed products and the margin for environmental improvement has been considerably reduced, making it more difficult for the product to achieve this.

4.3 Results Obtained

The introduction of ecodesign standard guidelines in the company's management system has made a series of contributions, among which special attention should be drawn to the obtaining of products with greater added value and the assistance provided in anticipating compliance with future environmental legislation, as was the case with the REACH regulations in the chemical sector or the *ErP* Directive in the electrical-electronic sector.

In addition, it has enabled them to gain greater power of innovation over the design and development of products and associated processes (supply chain, manufacturer, shipment, installation, operation and maintenance, etc.). Some companies highlight the introduction of LCT (*Life Cycle Thinking*) in the product design and development systems in organisations. In any event, practically all companies maintain that it provides another perspective of the process, enabling environmental impact to be improved—in some phases more than others, according to sector and product. Only in the case of the furniture sector would Ofita seem to distance itself and state that it has not detected any innovation as a result of ecodesign.

At the same time and despite the fact that it provides another perspective in creating new products, A&B Laboratorios and DTS OABE stress that ecodesign adoption also means to a certain extent less freedom when creating such new products.

All companies share the opinion that ecodesign is a competitive tool although, with the tools at their disposal, it is difficult to showcase it from a commercial standpoint—they highlight the major effort made in launching eco-designed products on the market. Yet markets have so far not taken into consideration the environmental impact of products in their purchasing criteria, and in any event, the companies interviewed point out that public administrative bodies need to promote measures to foster the development of *green* markets. Their role is an essential one. Despite being both extensive and restrictive, current legislation needs to improve and bring itself up-to-date with the state of business techniques and practice—companies unanimously suggest the development of legislation governing environmental matters. Prominence needs to be given to the most environmentally-efficient products or those that evidence the worst behaviour should be taxed accordingly—not to do so would create a situation of unfair competition, as many companies would make no effort to innovate their products. Rather, they manufacture less efficient ones that have greater environmental impact but at lower cost.

However, one direct consequence of applying ecodesign in practically all companies has been a slight increase in the cost of the product during the design phase for several reasons, which has in turn been reflected in a slight increase in its end price. As regards the latter, Inteman add that "offering it at the same price as its traditional equivalent would detract from the eco-designed product." It has greater added value and, broadly speaking, has helped to improve the product's quality. Despite this, operative costs are also higher in the case of all sectors, even where the standard has enabled management control of the product to be improved, and this is due to greater investment in time and resources during the product's design and development phase. In addition, materials and components are in general slightly more expensive, this being most apparent in the chemical and consumer goods sectors.

In any event, none of the companies subject to study say that they have measured the influence of adopting the standard either on general costs or on business results. Adopting the standard has not entailed any noteworthy reduction in costs, given that cost reduction has traditionally been a maxim in company activities—and it is precisely the pressure of costs that is a determining factor in the margin for environmental improvement of the product, due to a cost reduction policy for the product imposed by highly competitive markets.

Nonetheless, in the chemical, electrical-electronic and consumer goods sectors in particular, in many cases cost saving in some or other of the subsequent phases of the product's life cycle (shipment, use and management at the end of the life cycle) is greater than the increase in the cost of acquiring the eco-designed product over the non-eco-designed one.

Despite all the above, companies think the ecodesign certificate will help them to meet the requirements demanded by both public and private customers, and to ensure they have a privileged position in *green* purchasing processes. The certificate essentially provides credibility and companies estimate that they will have gained in terms of brand and company image with certification. This has enabled them to differ from the competition and gain certain consumer confidence.

In any event, the chemical and furniture sectors would seem to opt for product recognition or certification systems. This is a trend that appears to be prevailing on international markets even in cases—as they point out—where the large number of systems used to assess environmental impact of products, certificates, brands and environmentally-friendly labels confuse the consumer and make it difficult to compare environmental behaviour among products with similar features. Nonetheless, they advocate this form of recognition because it enables them to compete on the same level internationally as other products from the competition. In the case of the other industrial sectors, i.e. the consumer goods and electrical-electronic groups, companies would seem not to opt for this form of recognition, because either no environmentally-friendly label exists for their products or because, as they state at BSH, such recognition systems are very costly. In both cases, companies do not directly use ecodesign as a marketing argument because they are unable to provide the customer with 100 % reliable information about the product's environmental behaviour, and try to highlight graphically and intuitively the product's functional features and environmental improvements using labels on the product.

5 Discussion and Conclusions

Diverse motives behind encouraging industrial firms to adopt and certify ecodesign management systems exist. Among them, the growing interest on the market in more environmentally-sustainable products, the search for new innovation tools, and differentiation in terms of competitors are motives shared by all companies, although, they do not stand out as the main ones. In any event, in all cases, environmental awareness-raising of individuals from the organizations themselves would seem to be the common denominator that fosters the search for tools to improve the environmental behaviour of their activities and the results deriving from them.

As for the results obtained from the implementation of ISO 14006, the companies subjected to analysis point out a major lack of awareness on the market of the ecodesign standard and of the implications that the certificate entails for the organization and its products. The customer assesses mainly those features of the product that best meet their needs and considers environmental behaviour as an *extra*. The customer does not consider improvement in the product's environmental behaviour as a reason for purchasing more in itself. However, signs of an increasing concern on the part of consumers with climate protection are occurring; although it is also true to say that this is still too small to have any effect on so-called *green* markets. Meanwhile, at these times of crisis, the price is proving to be a more determining purchasing factor than ever.

Regardless of the effort made to highlight more environmentally-sustainable products, in all companies has been improved internal and external communication, owing partly to the importance given to this chapter in the standard and as many of the companies confirm, companies are still having problems with employees adopting the new strategies. This has translated into the design and setting in motion of courses of action and initiatives aimed at improving external communication with customers, suppliers, the sales network and other interest groups, and also at the heart of the organisation—their being convinced of the need on the part of the people within the organisation themselves to note the result of their efforts and internalise environmental improvement in the product as a constant feature of their activity.

The companies subject to study confess that such environmental improvement in the product is easy to manipulate—some results or other are obtained depending on the data base and/or software used to assess impact. It is also possible to take an initial reference using a severe environmental impact that may facilitate implementation of environmental improvements with minimal effort, and this situation raises doubts in the companies analysed in all sectors. Conversely, some companies from the chemical and consumer goods sectors warn that if the ecodesign standard were less flexible and contributed towards better development and precision, this would restrict company freedom and therefore creativity, thus making it difficult to use as an innovation tool.

They also maintain such flexibility can facilitate the standard's dissemination and spread the word about ecodesign internationally. It is precisely dissemination that is one of the areas about which all companies express doubts. They hope that in the future customers will ask for the ecodesign certificate just as they do now with other management certificates in accordance with standards, although they also fear this not to be the case insofar as there is an open debate about the possibility of being able to become certified in accordance with this, given that it has no certification purpose.

This circumstance and the fact that public administrative bodies are not carrying out more active work to develop green markets may affect dissemination of the ecodesign standard, even bearing in mind that ISO 14006 is an international standard. Despite this, companies appear satisfied with adoption and certification of the ecodesign standard, as it is a tool that improves and complements the suitable management of processes in the company by providing a very useful system that may help to streamline day-to-day procedures.

Finally, regarding the limitations of the research, it is necessary to mention those limitations of the case study methodology. A major limitation lies in the problems associated with the generalization of the results obtained through a limited number of cases, an aspect that has been criticized in the past, but the modern case study has overcome. The results of a case study do not allow for statistical generalizations, however, it is necessary talk about the analytical generalization, i.e. generalize a particular set of results of a case similar to others representing theoretical conditions.

Furthermore, Eisenhardt (1989) suggests that for analytical generalization, performing a cross-case with four to ten cases of study. Nevertheless, in the case study carried out in all sectors (except electrical-electronics) has not been able to have more than three cases. However, in this regard Eisenhardt (1989) also states it is necessary to limit the number of cases reaching a point of theoretical saturation. Other authors such as Yin (2009) note that the number of cases depends on the certainty desired on their performance and ability to predict similar results for the existence of predictable reasons. In the light of the results, in the present research, it has reached a point of theoretical saturation in the chemical and capital goods sectors. This means it will be necessary to seek further evidence to corroborate and appropriately generalize the results in other two sectors, furniture and electrical-electronics.

Therefore, the cases results could be generalized to other companies in the same sector or even some of the results could be generalized to the industry in general, taking into account the nuances and the conditions for doing so.

Another aspect that should be noted about the research topic addressed in this paper is that, since the ISO 14006 standard is in its early stages of diffusion, many issues still need to be undertaken and its analysis can be of great interest to the world academic and different interest groups that participate in the process of adoption and certification. Among them, attention should be drawn to the analysis of the keys to regional and inter-sector dissemination of the standard, the role that will be played by Ecodesign Management Systems for companies and its value on the market—bearing in mind that it has no certification purpose—and, especially the long-term influence of Ecodesign Management on financial results and company sustainability. Indeed, the number of certified companies grows gradually, allowing the combined application of quantitative and qualitative research methodologies that enrich and extend the existing knowledge of the phenomenon under study. Also, access to staff not involved in the adoption process will be a key issue in future research that it will be worked especially in order to obtain results that provide a more objective understanding of the phenomenon.

Acknowledgments This article is a result of a Research Group funded by the Basque Autonomous Government (Grupos de Investigación del Sistema Universitario Vasco; GIC12/158 - IT763/13).

References

Arana-Landín G, Cilleruelo E, Aldasoro JC (2012) ISO 14006. Experiencias previas de estudios de arquitectura que han adoptado el estándar de ecodiseño UNE 150301: 2003. Informes de la Construcción 64(527):319–330

Arana-Landin G, Heras-Saizarbitoria I (2011) Paving the way for the ISO 14006 ecodesign standard: an exploratory study in Spanish companies. J Cleaner Prod 19(9):1007–1015

Arana-Landin G, Heras-Saizarbitoria I, Cilleruelo-Carrasco E (2012) A case study of the adoption of a reference standard for ISO 14006 in the lift industry. Clean Technol Environ Policy 14 (4):641–649

Arana-Landin G, Landeta B, Ruiz de Arbulo P, Díaz de Basurto P (2013) Análisis de los efectos de la adopción de estándares de ecodiseño en el rendimiento empresarial de los estudios de arquitectura. DYNA 80:201–209

AENOR (2003) UNE150301: 2003. Gestión ambiental del proceso de diseño y desarrollo. Ecodiseño, Madrid

Bovea MD, Pérez-Belis V (2012) A taxonomy of ecodesign tools for integrating environmental requirements into the product design process. J Cleaner Prod 20(1):61–71

Chen Y, Okudan GE, Riley DR (2010) Sustainable performance criteria for construction method selection in concrete buildings. Autom Constr 19(2):235–244

Cluzel F (2012) Eco-design implementation for complex industrial system: from scenario-based LCA to the definition of an eco-innovative R&D projects portfolio (tesis doctoral) École Centrale des Arts et Manufactures "École Centrale Paris", Paris, France

Collado-Ruiz D, Ceca MJB, Ruescas-Nicolau A, Pacheco-Blanco B, Viñoles-Cebolla R (2008) Diseño para el fin de vida de pequeños electrodomesticos. Caso de estudio de pequeños estudios dentales. DYNA Ing e Industria 83(8):507–515

Comoglio C, Botta S (2012) The use of indicators and the role of environmental management systems for environmental performances improvement: a survey on ISO 14001 certified companies in the automotive sector. J Cleaner Prod 20(1):92–102

Deutz P, McGuire M, Neighbour G (2013) Eco-design practice in the context of a structured design process: an interdisciplinary empirical study of UK manufacturers. J Cleaner Prod 39:117–128

Eisenhardt KM (1989) Building theories from case study research. Acad Manage Rev 14(4):532–550

Fernández JM (2012) Ecodiseño en los sistemas de gestión ambiental según UNE-EN ISO 14006:2011. Aplicación a una empresa de base tecnológica. Dyna 88(5):514–517

Heras-Saizarbitoria I, Boiral O (2013) ISO 9001 and ISO 14001: towards a research agenda on management system standards. Int J Manage Rev 15(1):47–65

Hischier R, Baudin I (2010) LCA study of a plasma television device. Int J Cycle Assess 15:428–438

ISO (2004) ISO 14001:2004. Environmental management systems. Requirements with guidance for use, Geneva

ISO (2006a) ISO 14064-1:2006. Greenhouse gases. Part 1: specification with guidance at the organization level for quantification and reporting of greenhouse gas emissions and removals, Geneva

ISO (2006b) ISO 14064-2:2006. Greenhouse gases. Part 2: specification with guidance at the project level for quantification, monitoring and reporting of greenhouse gas emission reductions or removal enhancements, Geneva

ISO (2006c) ISO 14064-3:2006. Greenhouse gases. Part 3: specification with guidance for the validation and verification of greenhouse gas assertions, Geneva

ISO (2011a) ISO 14006:2011. Environmental management systems. Guidelines for incorporating ecodesign, Geneva

ISO (2011b) ISO 50001:2011. Energy management systems. Requirements with guidance for use, Geneva

ISO (2013a) ISO 14031:2013. Environmental management. Environmental performance evaluation. Guidelines, Geneva

ISO (2013b) 27001:2013. Information technology. Security techniques. Information security management systems. Requirements, Geneva

Justel-Lozano D (2008) Metodología para la eco-innovación en el diseño para el desensamblado de productos industriales. Universitat Jaume I, Castellón

Kara S, Honke I, Kaebernick H (2005) An integrated framework for implementing sustainable product development. Environmentally conscious design and inverse manufacturing. In: Fourth international symposium on eco design 2005. IEEE, pp 684–691

Knight P, Jenkins JO (2009) Adopting and applying eco-design techniques: a practitioners perspective. J Cleaner Prod 17(5):549–558

Landeta B, Arana G, Ruíz de Arbulo P, Díaz de Basurto P (2013) El estándar ISO14006: Una herramienta que ayuda a mejorar la competitividad de los estudios de arquitectura. Revista Ing de Construcción 28(2):155–171

Luttropp C, Lagersted J (2006) Ecodesign and the ten golden rules: generic advice for merging environmental aspects into product development. J Cleaner Prod 14(15):1396–1408

McKane A (2010) Thinking globally: how ISO 50001-energy management can make industrial energy efficiency standard practice. Lawrence Berkeley National Laboratory, California

Miles MB, Huberman AM (1994) Qualitative data analysis: an expanded sourcebook. Sage, California

Muñoz I, Gazulla C, Bala A, Puig R, Fullana P (2009) LCA and ecodesign in the toy industry: case study of a teddy bear incorporating electric and electronic components. Int J Cycle Assess 14:64–72

Pigosso DCA, Sousa SR (2011) Life cycle assessment (LCA): discussion on full-scale and simplified assessments to support the product development process. In 3rd international workshop: advances in cleaner production, São Paulo, Brazil

Rio M, Reyes T, Roucoules L (2013) Toward proactive (eco) design process: modeling information transformations among designers activities. J Cleaner Prod 39:105–116

Salomone R, Rupo D, Saija G (2013) Innovative environmental management tools for the agri-food chain. In: Product-oriented environmental management systems (POEMS). Springer, Netherlands, pp 3–25

Sanchez I, Pulido H, McNeil MA, Turiel I, Della Cava M (2007) Assessment of the Impacts of Standards and Labeling Programs in Mexico. Lawrence Berkeley National Laboratory, USA

Sanyé-Mengual E, Lozano RG, Farreny R, Oliver-Solà J, Gasol CM, Rieradevall J (2014) Introduction to the eco-design methodology and the role of product carbon footprint. In: Assessment of carbon footprint in different industrial sectors, vol 1. Springer, Singapore, pp 1–24

Savita KS, Dominic PD, Ramayah T (2012) Eco-design strategy among ISO 14001 certified manufacturing firms in Malaysia: green drivers and its relationship to performance outcomes. In: 2012 International conference on computer and information science (ICCIS), vol 1. IEEE, pp 154–159

Soosay CA, Hyland PW, Ferrer M (2008) Supply chain collaboration: capabilities for continuous innovation. Supply Chain Manage Int J 13(2):160–169

Stechemesser K, Guenther E (2012) Carbon accounting: a systematic literature review. J Cleaner Prod 36:17–38

Yin RK (2009) Case study research: design and methods. Sage, California

Early Adoption of ISO 50001 Standard: An Empirical Study

Maria Wulandari, Iker Laskurain, Martí Casadesús Fa and Iñaki Heras-Saizarbitoria

Abstract The adoption of international standards for Environmental Management Systems (EMSs) has grown significantly over the last years. Following this successful path some other management standards which deal with environmental and energy management issues have been launched as well. This is the case of ISO 50001, a certifiable international standard to adopt an Energy Management System (EnMS), which was launched in 2011 by the International Organization for Standardization (ISO). From then on, ISO 50001 certifications have experienced a huge growth: from 459 in 2011 to 1981 in 2012, with a total annual growth of 332 %. Regardless of the huge growth, within the scholarly literature only case-based studies have been published on the early adoption of this standard. In order to fulfill this gap this contribution summarizes some of the preliminary results of an empirical study carried out in Spain with the participation of a total of 57 early ISO 50001 certified organizations, aimed at analyzing the early adoption of ISO 50001 standard. The findings facilitate the characterization of ISO 50001 certified organizations and anticipate benefits of the adoption of the standard.

Keywords ISO 50001 · Energy management · Environmental management · Survey · Spain

---

M. Wulandari · M.C. Fa
Departament D'Organització, Gestió Empresarial I Disseny Del Producte, Escola Politècnica Superior, University of Girona, Campus de Montilivi, 17071 Girona, Spain
e-mail: wulan_gp05@yahoo.com

M.C. Fa
e-mail: marti.casadesus@udg.es

I. Laskurain · I. Heras-Saizarbitoria (✉)
Department of Management, University of the Basque Country UPV/EHU, Plaza Oñati 1, 20018 Donostia-San Sebastian, Spain
e-mail: iheras@ehu.es

I. Laskurain
e-mail: iker.laskurain@ehu.es

A. Chiarini (ed.), *Sustainable Operations Management*, Measuring Operations Performance, DOI 10.1007/978-3-319-14002-5_9

1 Introduction

Since it was created in 1987 the International Organization for Standardization (ISO) has launched various management systems standards. Known as meta-standards (Heras-Saizarbitoria and Boiral 2013a), they are in many respects different from the technical standards or regulations, which provide the requirements which certain products or processes need to meet. To name a few of the most well known meta-standards, for instance ISO 9000 family—quality management, ISO 14000 family—environmental management, ISO 26000—social responsibility and ISO 50001—Energy management standard. Although there are plenty meta-standards, this contribution is focused on one specific management system, namely ISO 50001 EnMS standard. Referring to the standardization of very diverse aspects of business activities, ISO 50001 standard contains similarly-structured models of management systems specific to a particular organizational function or stakeholder.

Energy is crucial for the survival of human being; therefore it is not a surprise that energy supply chain, from production to consumption is one of the most important activities of human life. With the rising of energy prices and the situation where energy is an organization's most significant environmental impact, the energy management has emerged as a crucial field. This is where ISO 50001 standard found its niche. ISO 14001 helps to identify all environmental impacts in the broad sense while ISO 50001 focuses on the continual improvement of energy performance, efficiency and consumption. ISO established ISO 50001 standard in order to support organizations to save money by using energy efficiently as well as helping to conserve resources and deal with climate change (Eccleston et al. 2011). ISO 50001 is expected to give a big impact in energy management in the present and in the future. "This standard established a framework, not only for industrial plants but also for commercial, institutional, governmental facilities; and entire organizations to manage energy. Targeting broad applicability across national economic sectors, it is estimated that the standard could influence up to 60 % of the world's energy use" (ISO 2011b).

Since the day it was established, ISO 50001 has been adopted in various industrial and commercial activities. The certification of ISO 50001 standards grows rapidly in only one year, from 459 on 2011 to 1981 on 2012, with total annual growth of 332 % (ISO 2013a). According to ISO 50001 certification world share, Europe has the highest world regional share followed by East Asia and Pacific. The ISO 50001 certification share of Europe 79 and 89 % for 2011 and 2012 respectively. Beside covering the highest ISO certification worldwide, Europe also showed an enormous growth of 383 % on ISO 50001 certification, followed by North America and East Asia and Pacific respectively. In Europe, the highest share of ISO 50001 comes from Germany with 1,115 certificates, followed by Spain, Denmark, Sweden, Italy, Romania, France, Ireland, Austria and United Kingdom respectively. In spite of the high quantity of certificates around the world, especially inside Europe, it is surprising to see that in the specialized literature only several case studies found (Chiu et al. 2012; Wessels 2011; Velázquez et al. 2013; Lambert 2013; Straughan 2013).

In order to fulfill this gap this contribution summarizes some of the preliminary results of an empirical study aimed at analyzing the early adoption of ISO 50001 standard. Specifically investigated in the survey were the motivations, the resources used, the difficulties faced, and the benefits achieved by the adoption of ISO 50001 energy management standard. In this chapter a fairly basic and descriptive analysis is provided, structured around the main topics of the questionnaire answered by the responding organizations. This rest of this contribution is arranged in the following manner. A detailed explanation regarding the structure of ISO 50001 is explained in the next section. Subsequently in the following section, the methodology of the survey and the survey profile are explained. Afterwards, the main results of the survey are presented in section number four. And finally, the overall results are recapitulated in the final section of conclusions.

2 Structure of ISO 50001

Before ISO 50001 was issued on June 2011, EN 16001:2009 has already been applied in Europe. EN 16001 was issued on July 2009 and it is now superseded by ISO 50001 and withdrawn on April 2012. The structure of ISO 50001 is designed according to other ISO management system standards, in particular ISO 9001 (Quality Management Systems) and ISO 14001 (Environmental Management Systems). Since all three management systems are based on the PDCA cycle, ISO 50001 can be integrated easily to these systems. The structure comparison between ISO 50001 with EN 16001 (Duglio 2011) and ISO 140001 is shown in Table 1. The novel parts in ISO 50001 standard compared with EN 16001 and ISO 14001 are shown in bold letters.

It can be seen that ISO 50001 and 14001 have similar structure of Scope and Normative references, while EN 16001 does not. However, between EN 16001 and ISO 50001, beyond some differences, which does not change substantially the contents, there are four steps of greater distinction:

1. The scope of ISO 50001 refers to continual improvement of energy performance, including energy efficiency, energy use and consumption while EN 16001 only focuses on energy efficiency.
2. The Management responsibility (in 4.2) and the subsequent subsections of the ISO 50001, absent in the EN 16001 and ISO 14001. ISO 50001 emphasizes the fundamental role of the so-called Top Management. It's a strategic actor, which defines policy, objectives and, consequently, allocates resources and defines operational roles (Duglio 2011).
3. The third aspect is present in the Plan phase where ISO 50001 adds some concepts (in 4.4.3—Energy Review, 4.4.4—Energy Baseline and 4.4.5—Energy performance indicators). Coming to the second aspect, in the Plan phase

Table 1 Structure comparison of ISO 50001 with EN 16001 and ISO 14001

Index	EN 16001	Index	ISO 50001	Index	ISO 14001
1	Scope	1	Scope	1	Scope
		2	Normative references[a]	2	Normative references
2	Terms and definitions	3	Terms and definitions	3	Terms and definitions
3	Energy management system requirements	4	Energy management system requirements	4	Environmental management system requirements
3.1	General requirements	4.1	General requirements	4.1	General requirements
		4.2	Management responsibility1		
		4.2.1	Top management a		
		4.2.2	Management representative1		
3.2	Energy policy	4.3	Energy policy	4.2	Environmental policy
3.3	Planning	4.4	Energy planning	4.3	Planning
3.3.1	Identification and review of energy aspects	4.4.1	General requirements a	4.3.1	Environmental aspects
3.3.2	Legal obligation and other requirements	4.4.2	Legal and other requirements	4.3.2	Legal and other environmental requirements
3.3.3	Energy objective, targets and programme(s)	4.4.3	Energy review a	4.3.3	Objectives, targets and programme(s)
		4.4.4	Energy baseline a		
		4.4.5	Energy performance a indicators		
		4.4.6	Energy objectives, energy targets and energy management action plans		
3.4	Implementation and operation	4.5	Implementation and operation	4.4	Implementation and operation
3.4.1	Resources, roles, responsibility and authority	4.5.1	General	4.4.1	Environmental aspects
3.4.2	Awareness, training and competence	4.5.2	Competence, training and awareness	4.4.2	Legal and other environmental requirements

(continued)

Table 1 (continued)

Index	EN 16001	Index	ISO 50001	Index	ISO 14001
3.4.3	Communication	4.5.3	Communication	4.4.3	Communication
3.4.4	Energy management system documentation	4.5.4	Documentation	4.4.4	Documentation
3.4.5	Control of documents	4.5.5	Operational control	4.4.5	Control of documents
3.4.6	Operational control	4.5.6	Design [a]	4.4.6	Operational control
		4.5.7	Procurement of energy services, products, equipment and energy [a]	4.4.7	Emergency preparedness and response
3.5	Checking	4.6	Checking	4.5	Checking and corrective action
3.5.1	Monitoring and measurement	4.6.1	Monitoring, measurement and analysis	4.5.1	Monitoring and measurement
3.5.2	Evaluation of compliance	4.6.2	Evaluation of legal requirements and other requirements	4.5.2	Evaluation of compliance
3.5.3	Nonconformity, corrective action and preventive action	4.6.3	Internal audit of the EnMS	4.5.3	Nonconformity, corrective and preventive actions
3.5.4	Control of records	4.6.4	Nonconformities, correction, corrective, and preventive action	4.5.4	Records
3.5.5	Internal audit of the energy management system	4.6.5	Control of records	4.5.5	Internal audit
3.6	Review of the energy management system by top management	4.7	Management review	4.6	Management review
3.6.1	General	4.7.1	General		
3.6.2	Inputs to management review	4.6.2	Input to management review		
3.6.3	Outputs from management review	4.6.3	Output from management review		

[a] Novel parts of ISO 50001 Standard
Source put together by the authors based on the ISO standards

it's interesting to analyze the concept of energy analysis (4.4.3—Energy Review). A section is entirely dedicated to this concept because thanks to it the organization should establish reference energy data (4.4.4—Energy baseline).

The span of time is decided by the organization itself and the purpose is to be a basis for comparison of changes in the organization's energy performance, measured through appropriate indicators (4.4.5—Energy Performance Indicators).

4. Finally, the Do phase of ISO 50001 system has been added two new paragraphs (4.5.6—Design and 4.5.7—Procurement of energy services, products, equipment and energy). Regarding the latter ones, the methodologies are not indicated for the selection of indicators, but the single organization can define method that will be assessed during the visit of the certification body.

In order to provide compatibility and integration opportunity between different ISO meta-standards, ISO 50001 uses the same method of continual improvement Plan-Do-Check-Act cycle as employed in ISO 14001 and ISO 90001 (Fig. 1).

There are several activities that need to be conducted to implement ISO 50001. The activity list of Plan-Do-Check-Act cycle of ISO 50001 is shown in Table 2 below. The novel activities in ISO 50001 are shown in bold.

The novel parts of ISO 50001 are located in Energy Planning and the Implementation and Operation. Energy planning at minimum includes the specific processes to improve energy performance. The schema of energy planning process according to ISO (2011a) is shown in Fig. 2.

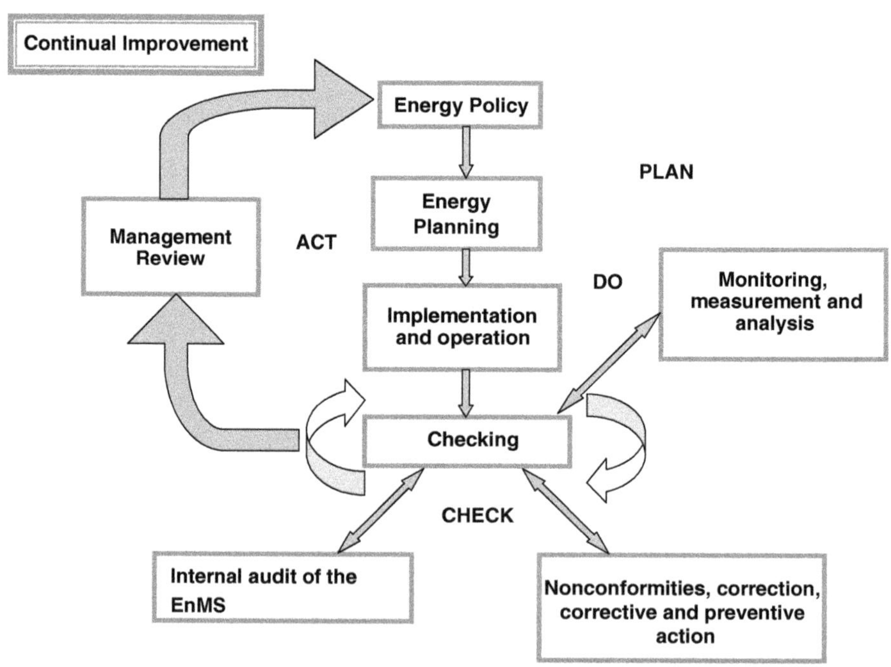

Fig. 1 Plan-Do-Check-Act Cycle of ISO 50001

Table 2 Activity List of ISO 50001

No	Cont. improvement	Activities	Energy management document output	Person in charge
1	PLAN	Establish energy policy	Energy policy	Top management
2		Appoint energy manager	–	Top management
3		Form energy management team	–	Energy Manager
4		Conduct energy review[a]	Energy review	Energy manager and energy management team
5		Establish energy baseline	Energy baseline	
6		Establish energy performance indicators (EnPIs)	EnPIs	
7		Establish energy objectives and targets	Objectives and targets	
8		Establish and action plans	Action plans	
9	DO	Ensure competent, training, awareness	–	
10		Communicates internally	–	
11		Documentation	EnMS documentation	
12		Operational Control	–	
13		Design	–	
14		Procurement energy services, products and equipments	–	
15	CHECK	Monitor and measure processes and key characteristics	–	
16		Evaluation of compliance with legal requirements and other requirements	–	
17		Internal audit of the EnMS	–	
18		Review and make corrections, corrective action, and preventive action.	–	
19		Establish and maintain records	–	
20		EnMS management review	Changes of EnMS documentation	Top Management
21	ACT	Take actions to continually improve energy performance		Energy manager and energy management team

[a] Novel activities of ISO 50001 Standard
Source put together by the authors

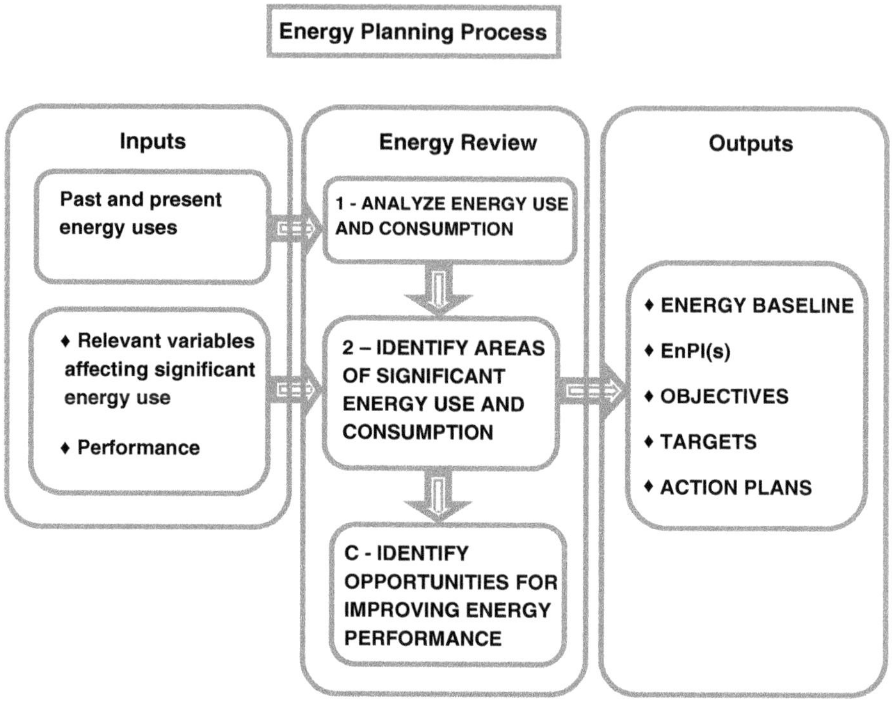

Fig. 2 Energy planning process concept diagram

- Energy review
  Energy review is the determination of the organization's energy performance based on data and other information, leading to identification of opportunities for improvement (ISO 2011a). It is a review of the organization's processes which affect energy use and consumption.
- Energy baseline
  Energy baseline is quantitative reference(s) providing a basis for comparison of energy performance (ISO 2011a). While according to Reichl and Kollmann (2011) the energy baseline is the energy consumption that would have occurred if no direct measures had been taken to influence energy consumption. The changes in energy performance shall be measured against the energy baseline (ISO 2011b).
- Energy Performance Indicators (EnPI)
  Energy performance indicator is quantitative value or measure of energy performance, as defined by the organization (ISO 2011a). It should be identified to assess energy performance and to subsequently evaluate progress towards objectives and targets. It could be expressed as a simple metric, ratio or a more complex model, depending on the organization need.

- Objectives and Targets
  Energy objective is specified outcome or achievement set to meet the organization's energy policy related to improved energy performance (ISO 2011a). Energy target is detailed and quantifiable energy performance requirement, applicable to the organization or parts thereof, that arises from the energy objective and that needs to be set and met in order to achieve this objective (ISO 2011a). Energy objectives and targets shall be established and implemented at the relevant processes and facilities within the organization inside the established time frames.
- Action plans
  According to ISO (2011a), the action plans shall include:
  - designation of responsibility;
  - the means and time frame by which individual targets are to be achieved;
  - a statement of the method by which an improvement in energy performance shall be verified;
  - a statement of the method of verifying the results.

After the energy action plan is prepared, it must be communicated to all the persons in charge in its implementation.

Those documents are to be developed by energy manager and energy management team. The energy management team must consist of professional from various technologies and discipline (Eccleston et al. 2011).

Regarding the other novel part, Implementation and Operation, in this part, the organization needs to conduct the entire plan that has been developed by energy manager and energy management team. ISO (2011b) requires an organization to:

- ensure that any person(s) working for or on its behalf, related to significant energy uses, are competent;
- provide training to meet these needs;
- communicate internally with regard to its energy performance and EnMS;
- establish, implement and maintain information;
- identify the operations and maintenance activities which are related to significant energy uses;
- consider energy performance improvement opportunities and operational control in the design of new, modified and renovated facilities, equipment, systems and processes;
- establish and implement the criteria for assessing energy use, consumption and efficiency over the planned or expected operating lifetime when procuring energy using products, equipment and services.

3 Methodology of the Survey

In order to investigate the motivations and benefits of organizations who have adopted ISO 50001 standard a survey was planned. A questionnaire was designed on the basis of a comprehensive literature review. First, it was conducted a literature review about ISO 50001, the search resulted in several case studies (Chiu et al. 2012; Wessels 2011; Velázquez et al. 2013; Lambert 2013; Straughan 2013), with no empirical studies found. Afterward, as a basic reference, it is considered the researches about ISO 9001 and ISO 14001 (Corbett et al. 2002; Liyin et al. 2006; Williams 2004; Gavronski et al. 2008; Karapetrovic et al. 2006; De Oliveira et al. 2010; Psomas et al. 2011) since it is logical to believe that the adoption of ISO 50001 will follow the same reasoning valid for the adoption of ISO 9001 and ISO 14001. Based on the basic reference, it was formulated the concepts and the various variables affecting it. In this way, it was established the relation between questions and analyzing categories in respect with motivations, implementations and benefits of the early adoption of ISO 50001. The literatures referring to similar and identical concepts were aggregated, generating a list of concepts and references. The survey contained four sections: (1)—organization data, (2)—organization input prior, (3)—implementation process of ISO 50001, and (4)—output of ISO 50001 adoptions. Each section has different content, as shown more detail in Tables 3 and 4.

A total of six pages of questions were employed using a combination of the one to five Likert scale and open-ended answers. The Likert scale provides five alternatives with different degrees of agreement: (a) completely agree, (b) partially agree, (c) do not agree nor disagree, (d) partially disagree, and (e) completely disagree. The survey was disseminated through internet (Survey Monkey) and through paper questionnaire to 87 organizations of the total 120 organizations based

Table 3 ISO 5000 survey sections

Chapter	Contents
1. Organization data	Organization classification
	International scope
2. Input	Motivations
3. Implementation	Commitment leadership
	Human resource
	Other resources
	Time and cost
	Difficulties
	Integration
4. Output	Operational benefits
	Financial benefits
	Innovation

Source put together by the authors

Table 4 Profile of the ISO 50001 studied organizations

Study date	October 2013–January 2014
Study population	120 ISO 50001:2011 certified organizations based in Spain in November 2013
Study sample	87 organizations
Number of responses	57 organizations
Response rate	65 %

Source put together by the authors

in Spain that were registered to ISO 50001. The envelope and the cover letter were addressed to the responsible person for energy and environment or quality of the company. The survey was carried out from October 2013 to January 2014.

4 Results

This section provides a very short discussion of the main results of the empirical work that we carried out. Regarding ISO 50001 standard in organization stemming from the 2014 survey (for a more detailed analysis of the survey can be found in Wulandari et al. 2014). The first factor relates to the motivations for the adoption of the ISO 50001:2011 standard. Afterward, an analysis of the implementation of this process, specifically the commitment leadership, human resource, time and cost, difficulties and integration with other standards, are illustrated. And finally, an analysis of the outputs of this process, in term of operational benefits including energy saving and environmental benefits, and financial benefits are described.

From an empirical perspective Bansal and Hunter (2003) propose three types of motive that lead organizations to implement ISO 14001: competitiveness, legitimation, and ecological responsibility. Similarly, Neumayer and Perkins (2005) underlined two main sources of motivation that lead organizations to implement ISO 14001: internal motives related to efficiency and, on the other hand, external motives related to the social pressure exerted by different agents to persuade company managers to adopt certain practices. In the same vein, Boiral and Roy (2007) has also stressed that the adoption of ISO 14001 can be driven by external pressures and the search for organizational legitimacy in the eyes of various stakeholders (e.g., clients, public authorities, environmental groups), or by internal motivations to improve environmental practices. While González-Benito and González-Benito (2005) differentiated the following four drivers for the adoption of EMSs: operational competitive motivations (costs, productivity), commercial competitive motivations (market, image, customers), ethical motivations, and relational motivations (regulators, local organizations). Figure 3 shows the motivation to adopt ISO 50001 and its importance attached to each, demonstrated by the answers obtained from Part II of the questionnaire “Motivation of ISO 50001”.

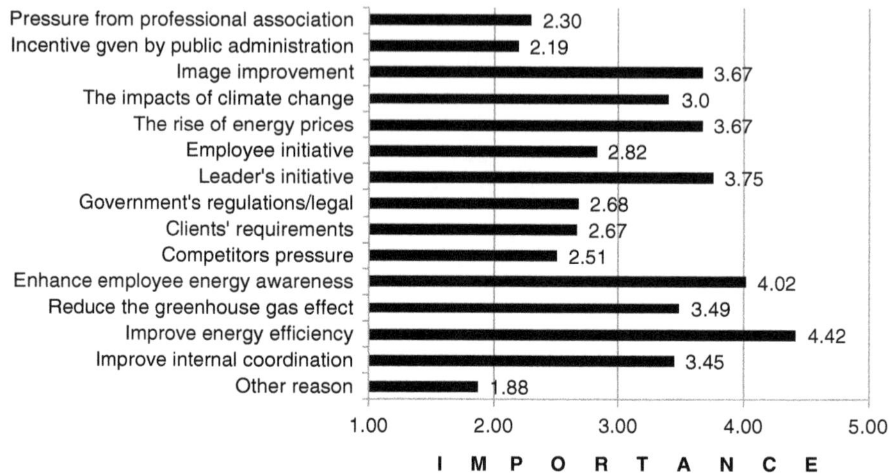

Fig. 3 Motivation for adopting ISO 50001 standard *Source* put together by the authors

By a long run, the most important factor to adopt ISO 50001 was the improvement of energy efficiency, with the median importance level of 4.42 out of 5. The second most important motivation was an issue related with enhancing energy awareness between employees, with the median of 4.02. These two are followed by leader's initiative (3.75), image improvement (3.67) and the rise of energy prices (3.67). And it seems that the drivers such as clients' requirements (2.67), government's regulations/legal (2.68), competitor's pressure (2.51), pressure from professional association (2.30) and incentive given by public administration (2.19) were not deemed as essential as the previously mentioned drivers.

In Fig. 4 the difficulties perceived in ISO 50001 adoption. The most evident difficulty perceived is the necessity of "Continuous measurement instruments" with

Fig. 4 ISO 50001 implementation difficulties *Source* put together by the authors

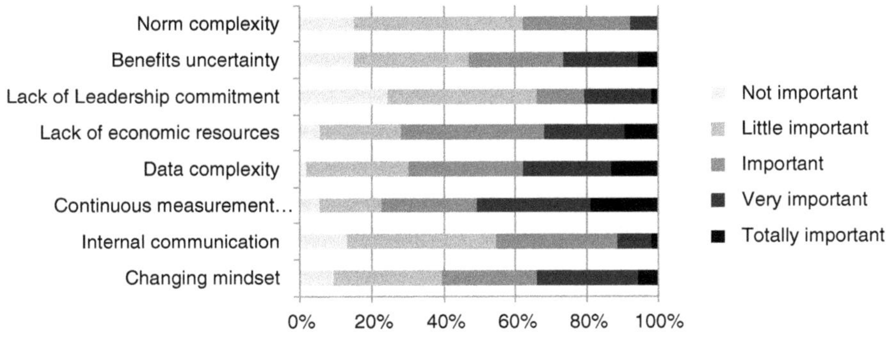

Fig. 5 Distribution of ISO 50001 implementation difficulties *Source* put together by the authors

the median of 3.42, while the second most important is "Data complexity" (3.19). They are followed by "Lack of economic resources" (3.08), "Changing Mindset" (2.91), "Benefits uncertainty" (2.70), "Internal communication" (2.45), "Lack of leadership commitment" (2.32).

Regarding the main obstacles or pitfalls Fig. 5 illustrates the distribution of the difficulties in ISO 50001 adoption. The higher degree of importance is symbolized with darker color; with the darkest indicate the most important difficulties. Seeing the little distribution of the dark color, it can be concluded that the adoption of ISO 50001 does not pose high difficulties. "Norm complexity" seems to be the least difficulty faced during the adoption, while "Continuous measurement instruments" is the highest difficulty faced by certified organizations. "Norm complexity" is perceived as the least difficulty because the parts and structure of ISO 50001 is similar with previous meta-standards, namely quality (ISO 9001), environment (ISO 14001), environment technology service (ISO 20000), corporate social responsibility (ISO 26000), information security (ISO 27000) and supply chain security (ISO 280001). And in line with Wessels (2011), the necessity of "Continuous measurement instruments" is the most difficult challenge in adopting ISO 50001.

Figure 6 illustrated the operational benefits obtained after adopting ISO 50001 and its importance attached to each. The most significant benefit resulted from the adoption of ISO 50001 is "Energy saving" with the median importance level of 4.43. This finding is in line with the study of Psomas et al. (2011) in Greek organizations with ISO 14001. It was shown that "Decrease of energy consumption" is among the benefits of higher importance after the ISO 14001 adoption. The following benefits are environmentally-related, namely "Improve environmental performance" with the median level of 4.02, increase of energy and environmental awareness with median of 3.62. The other benefits are related with day to day operation such as "Process optimization"(3.51), "Overall productivity" (3.06) and "Plant safety" (2.49). Unsurprisingly, the benefits of lowest importance are product related, namely "Improve product quality" (2.45) and "Improve product performance" (2.40).

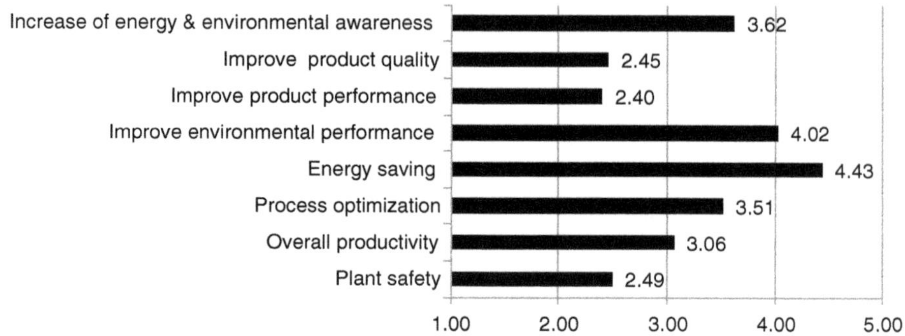

Fig. 6 ISO 50001 operational benefits *Source* put together by the authors

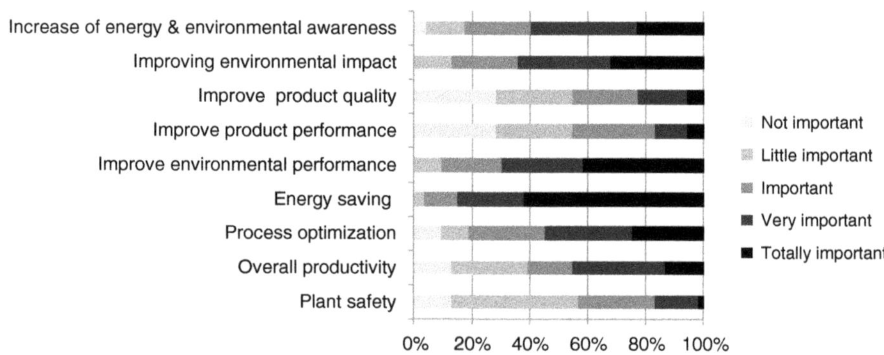

Fig. 7 Distribution of ISO 50001 operational benefits *Source* put together by the authors

Figure 7 illustrates the distribution of the operational benefits for adopting ISO 50001. The higher degree of importance is symbolized with darker color; with the darkest indicate the most important benefits. It is can be concluded easily that "Energy saving" is the most important benefit while "Improve product performance" is the least important operational benefits. As the highest importance of operational benefits is "Energy saving", it is interesting to identify how big is the quantity of energy saving encountered after the adoption of ISO 50001 shown in Fig. 8.

Half of the organizations adopting ISO 50001 achieved 1–5 % energy saving while the rest of the organizations achieve higher quantity of energy saving. For instance 29 % of the organizations achieved 6–10 % energy saving, 11 % achieved 11–15 % and 10 % achieved 16–20 %. This is in line with a study from United States Department of Energy of Superior Energy Performance (SEP) certification resulting that energy saving in the first year after SEP training is 3.8 % and 10.1 % in the first half of the second year (Therkelsen et al. 2013). The quantity of energy saving gained is ranging from 1–5 % and 6–10 % in accordance with energy

Fig. 8 Quantity of energy saving

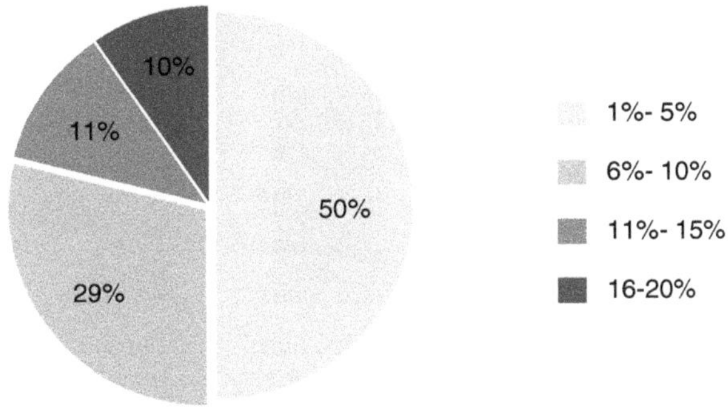

management "rule of thumb" from Capehart et al. (2006) saying that typical energy savings of first year energy management program is around 5–10 %.

The financial benefits vs adoption cost after the adoption of ISO 50001 is presented in Fig. 9. As can be seen, majority (51 %) responded that the benefit is bigger than the cost, 16 % responded that the benefit and the cost are similar, while 17 % responded that the adoption cost are bigger than the benefit and 16 % do not know yet.

However, the finding from the present study revealed that the response is not extremely high, only 51 %. A research by Carbon Trust Advisory Services (2013) shown that the average payback (a point in time when the cost spent and the profit gained is equal) of energy efficiency investment is one until 5 years. Considering that ISO 50001 was launched on July 2011, thus the adoption time in organizations so far ranges from 1–3 years, thus it is logical that some of the organizations which adopted ISO 50001 have not reached their payback. In the same vein, Psomas et al. (2011) in Greek organizations with ISO 14001 adoption pointed that it takes an extensive period, which seems more than five years, of ISO 14001 adoption to realize a better environmental performance. Therkelsen et al. (2013) constructed an arithmetic model based on case studies in nine facilities adopting SEP certification stating that SEP participation is expected to have less than 2-year payback for facilities with an annual energy consumption level greater than 0.27 TBtu, and shorter payback time with greater annual energy consumption.

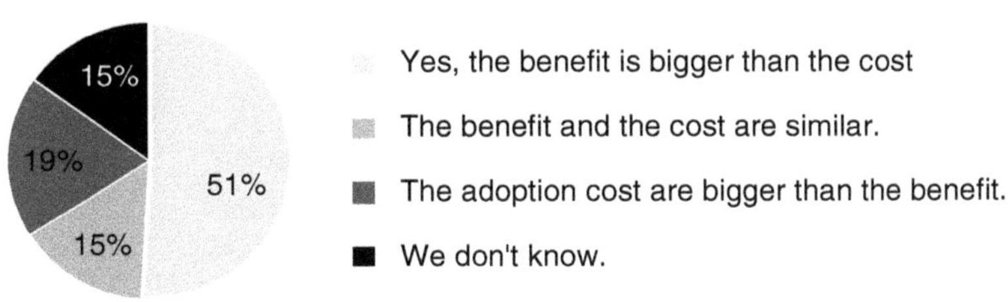

Fig. 9 Financial benefits vs cost after the adoption of ISO 50001 *Source* put together by the authors.

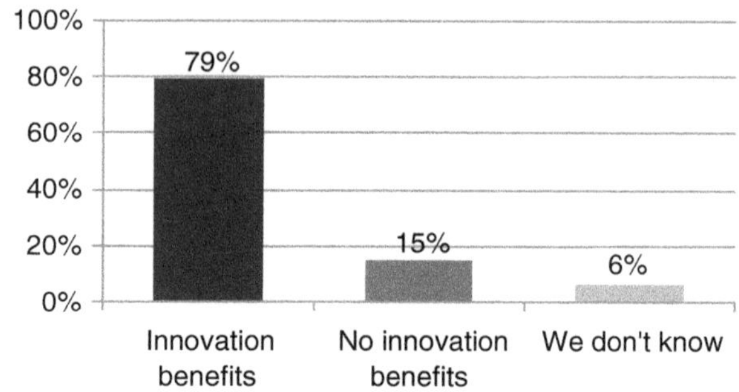

Fig. 10 Organization responds of innovation benefits for adopting ISO 50001 *Source* put together by the authors.

Figure 10 demonstrated that besides giving operational benefits and financial benefits, ISO 50001 also gives innovation benefits. Majority of organizations adopting ISO 50001 responded "Yes" that they reaped innovation benefits for adopting ISO 50001, while 15 % responded "No" and 6 % responded 'We do not know".

The highest innovation benefit for adopting ISO 50001 is "Innovative strategies" with the median importance level of 3.86. It is closely followed by the second highest innovation benefit namely "Process innovation" with the median of 3.58. They are followed by "Primary material innovation" (2.33) and "Final product innovation" (2.19). The finding is in line with Halila (2007) in Sweden SMEs with ISO 14001. He mentioned that corporate environmental strategy has progressed from compliance (i.e. reacting to environmental regulations) to innovative and proactive strategies (i.e. doing more than required by the regulatory authorities) Fig. 11.

Finally, Fig. 12 shows general satisfaction with the adoption of ISO 50001, which as well situated in the last part of the questionnaire to rate the satisfaction. The respondents were given multiple choices between "Very Satisfied" scored 5, until "Very unsatisfied" scored 1. The median importance level is 4.36 out of 5, significance "Very satisfied". The majority of respondents (55 %) respond "Satisfied" while 42 % respond "Very satisfied", leaving only 2 % responded "Little satisfied", 2 % of "Less satisfied" and 0 % of "Very unsatisfied".

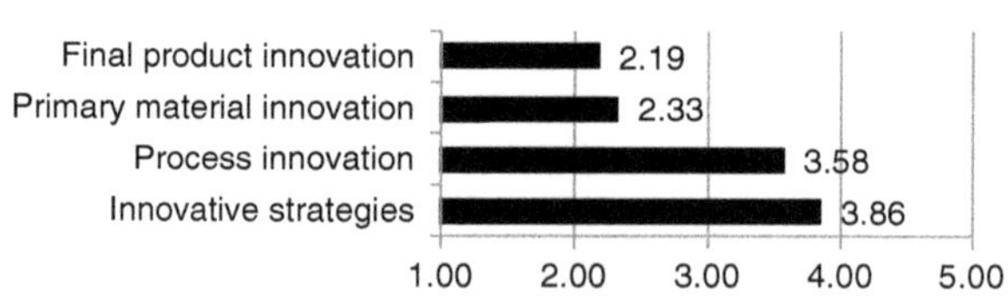

Fig. 11 Innovation benefits obtained for adopting ISO 50001 *Source* put together by the authors.

Fig. 12 ISO 50001 general satisfaction *Source* put together by the authors

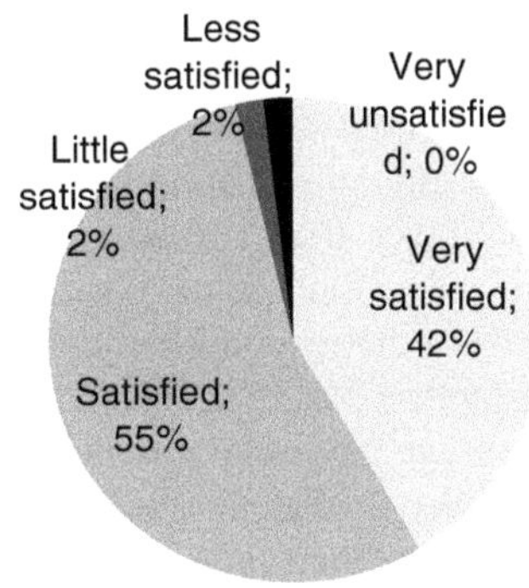

5 Conclusions

Our findings based on the survey have revealed the motivations and benefits of the early adoption of ISO 50001. The three most significant motivations to adopt ISO 50001 were the internal drivers such as "improving energy efficiency", "enhancing energy awareness between employees" and "leader's initiative" respectively. In our study it was concluded that internal drivers rather than external drivers motivate the implementation of ISO 50001. Based on the result obtained from the questionnaire, the adoption of ISO 50001 does not pose high difficulties. "Norm complexity" seems to be the least difficulty faced during the adoption, while "Continuous measurement instruments" is the highest difficulty faced by certified organizations. "Norm complexity" is perceived as the least difficulty because the parts and structure of ISO 50001 is similar with previous meta-standards. And in line with Wessels (2011), the necessity of "Continuous measurement instruments" is the most difficult challenge in adopting ISO 50001. According to the survey summarized in this contribution, the process benefits among the organizations adopting ISO 50001 are "energy saving", "environmental performance improvement" and "increase of energy and environmental awareness" are most frequently mentioned as the main benefits to adopt ISO 50001 standard compared to product benefits of "product performance" and "product quality". This confirmed that the adoption of ISO 50001 gives process benefits rather than product benefits.

This is in line with the study of De Oliveira et al. (2010) regarding ISO 14001 adoption at industries in São Paulo, Brazil that ISO 14001 adoption contributes especially to the reduction of energy consumption in the production process. In addition, the finding from the study of Psomas et al. (2011) in Greek companies with ISO 14001 is in line with findings from present survey. It was shown that "decrease of energy consumption" is among the benefits of higher importance after the ISO 14001 adoption.

Moreover, the majority of energy saving gained is ranging from 1–5 % and 6–10 % in accordance with energy management "rule of thumb" from Capehart et al. (2006) saying that typical energy savings of first year energy management program is around 5–10 %. However it is interesting to note that several respondents have gained 11–15 % and 16–20 % energy saving. This is possibly because they are the first organizations/companies who applied ISO 50001 or they have energy management program already in place for long. It is found that half of the respondents opinionated that the direct financial benefits of ISO 50001 adoption is bigger than the adoption cost. However, the finding from the present study revealed that the response is not extremely high and there exist a small finding where the respondents said that financial benefits are equal or even lower than the adoption cost. A research by Carbon Trust Advisory Services (2013) shown that the average payback (a point in time when the cost spent and the profit gained is equal) of energy efficiency investment is one until five years. Considering that ISO 50001 was launched on July 2011, thus the adoption time in organizations so far ranges from 1–3 years, thus it is logical that some of the organizations which adopted ISO 50001 have not reached their payback. In the same vein, Psomas et al. (2011) in Greek companies with ISO 14001 adoption pointed that it takes an extensive period, which seems more than five years, of ISO 14001 adoption to realize a better environmental performance. Furthermore, it is found some percentages of the respondents who are not familiar if the financial benefits are bigger than the adoption cost or not. This is perhaps due to the complexity of organization/company where some people only understand some part of the organizational activities and also due to lack of coordination inside organization/company.

The implications of these findings will be of great interest to the main stakeholders involved in the adoption of ISO 50001, especially for managers, consultants, certified bodies and the policy makers. In the specialized literature of the field internal factors of motivations to adopt management systems have been connected to higher internalization of the standard, or, in other words, to a more substantive adoption of meta-standards, while external drivers are more related to symbolic adoptions (e.g., Heras-Saizarbitoria et al. 2011; Neugebauer 2012; Heras-Saizarbitoria and Boiral 2013b), in other words, aimed at improving corporate image and legitimacy among stakeholders (Boiral 2007; Boiral 2011, 2012; Heras-Saizarbitoria and Boiral 2013a). Then, this evidence leads us to conclude that the adoption of ISO 50001—a more technical and specific meta-standard, which has, so far, a lower brand value for adopting organizations compared to a more disseminated and popular standard such as ISO 14001—, could be more aimed at gaining specific energy efficiency performance improvements than at other objectives.

Apart from the descriptive analysis, the success of this study is shown by the high response rate obtained in the survey and case studies of ISO 50001. The questionnaire was disseminated to 87 organizations of the total 120 ISO 50001 certified Spanish organizations (ISO 2013a), out of which 57 responded. The highest appreciation is given to 57 organizations who have expressed their experience and opinion in survey study.

Finally, possible limitations to this pioneering survey and possible avenues for future research should be mentioned. The methodology used to obtain the quantitative information on the adoption of ISO 50001, as is based on the perceptions of specialized managers in charge of the process, could suffer from social desirability and other related bias (Heras-Saizarbitoria and Boiral 2013a). Similarly, as underlined by Boiral and Roy (2007), reverse causality bias could be a problem, since the outcomes of ISO 50001 adoption could influence the perception of its drivers. Besides, this study is a static study measured in one point in time from one meta-standard that is still very young and thus the reality of ISO 50001 adoption may change in the future following the maturation of the standard itself. Also, although the main characteristics of the process of adoption of meta-standards does not differ much from one region to another, since the organizational field in which these kind of standards have been disseminated is a global one, specific conditions in other countries and regions may alter the findings (Heras-Saizarbitoria and Boiral 2013a). In any case, this explorative study serves as a stepping-stone for a more detailed and more comprehensive study to be conducted in the future. Future surveys might, on the one hand, collect, analyze and triangulate information from various information sources and on the other hand, collect sample from various countries to analyze the influence of the geographical variable.

Acknowledgments The authors wish to deeply thank Universitat de Girona and Erasmus Mundus Action 2 Techno Scholarship for funding this research. Likewise, this contribution is a result of a Research Group funded by the Basque Autonomous Government (*Grupos de investigación del sistema universitario vasco*; IT763-13 /GIC12-158).

References

Bansal P, Hunter T (2003) Strategic explanations for the early adoption of ISO 14001. J Bus Ethics 46(3):289–299

Boiral O (2007) Corporate greening through ISO 14001: a rational myth? Organ Sci 18(1):127–146

Boiral O, Roy MJ (2007) ISO 9000: integration rationales and organizational impacts. Int J Oper Prod Manage 27(2):226–247

Boiral O (2011) Managing with ISO systems: lessons from practice. Long Range Plan 44(3):197–220

Boiral O (2012) ISO certificates as organizational degrees? Beyond the rational myths of the certification process. Organ Stud 33(5–6):633–654

Capehart BL, Turner WC, Kennedy WJ (2006) Guide to energy management. The Fairmont Press, Inc., Georgia

Carbon Trust Advisory Services (2013) The business of energy efficiency. http://www.carbontrust.com/media/135418/cta001-business-of-energy-efficiency.pdf. Accesed 6 June 2013

Chiu T, Lo S, Tsai Y (2012) Establishing an integration-energy-practice model for improving energy performance indicators in ISO 50001 standards. Energies 5(12):5324–5339

Corbett CJ, Montes MJ, Kirsch DA, Alvarez-Gil MJ (2002) Does ISO 9000 certification pay? ISO Manage Syst 2(4):31–40

De Oliveira OJ, Serra JR, Salgado MH (2010) Does ISO 14001 work in Brazil? J Cleaner Prod 18 (18):1797–1806

Duglio S (2011) Energy management systems: from EN 16001 to ISO 50001. Management 4 (4.2):1

Eccleston CH, March F, Cohen T (2011) Inside energy: developing and managing an ISO 50001 standard. CRC Press LLC, NY

Gavronski I, Ferrer G, Paiva EL (2008) ISO 14001 certification in Brazil: motivations and benefits. J Cleaner Prod 16(1):87–94

González-Benito J, González-Benito O (2005) An analysis of the relationship between environmental motivations and ISO14001 certification. Brit J Manage 16(2):133–148

Halila F (2007) Networks as a means of supporting the adoption of organizational innovations in SMEs: the case of Environmental Management Systems (EMSs) based on ISO 14001. Corp Soc Responsib Environ Manage 14(3):167–181

Heras-Saizarbitoria I, Landín GA, Molina-Azorín JF (2011) Do drivers matter for the benefits of ISO 14001? Int J Oper Prod Manage 31(2):192–216

Heras-Saizarbitoria I, Boiral O (2013a) ISO 9001 and ISO 14001: towards a research agenda on management system standards. Int J Manage Rev 15(1):47–65

Heras-Saizarbitoria I, Boiral O 2013b Symbolic adoption of ISO 9000 in small and medium-sized enterprises: the role of internal contingencies. Int Small Bus J. doi:10.1177/0266242613495748

ISO (2011a) ISO 50001 Energy management systems—requirements with guidance for use. Energy review, international organization for standardization, Geneva, Switzerland 7–8. 18 June

ISO (2011b) Win the energy challenge with ISO 50001. ISO 50001:2011. Geneva, Switzerland

ISO (2013a) ISO Survey 2012. http://www.iso.org/iso/home/standards/certification/iso-survey.htm Accessed 10 Nov 2013

Karapetrovic S, Casadesús M, Heras-Saizarbitoria I (2006) Dynamics and integration of standardized management systems: an empirical study. Documenta Universitaria, Girona

Lambert G 2013 Never tire of saving energy! ISO 50001 is a win-win for Bridgestone. http://www.iso.org/iso/home/news_index/news_archive/news.htm?refid=Ref1723 Accessed 2 June 2013

Liyin S, Hong Y, Griffith A (2006) Improving environmental performance by means of empowerment of contractors. Manage Environ Qual Int J 17(3):242–257

Neugebauer F (2012) EMAS and ISO 14001 in the German industry–complements or substitutes? J Cleaner Prod 2(37):249–256

Neumayer E, Perkins R (2005) Uneven geographies of organizational practice: explaining the cross-national transfer and diffusion of ISO 9000. Econ Geogr 81(3):237–259

Psomas EL, Fotopoulos CV, Kafetzopoulos DP (2011) Motives, difficulties and benefits in implementing the ISO 14001 environmental management system. Manage Environ Qual Int J 22(4):502–521

Reichl J, Kollmann A (2011) The baseline in bottom-up energy efficiency and saving calculations–a concept for its formalisation and a discussion of relevant options. Appl Energy 88(2):422–431

Straughan M (2013) Bentley first with ISO 50001 car maker implements energy management standard. http://www.iso.org/iso/home/news_index/news_archive/news.htm?Refid=Ref1523 Accessed 2 June 2013

Therkelsen P, McKane A, Sabouni R, Evans T, Scheihing P (2013) Assessing the costs and benefits of the superior energy performance program. In: Proceedings of the 2013 ACEEE summer study on energy efficiency in industry, American Council for an Energy-Efficient Economy, Niagra Falls, NY

Velázquez D, González-Falcón R, Pérez-Lombard L, Marina Gallego L, Monedero I, Biscarri F (2013) Development of an EnMS for a naphtha reforming plant: a data mining approach. Energy Convers Manage 67:217–225

Wessels A (2011) EnMS adoption at Toyota SA. In: Proceedings of the 8th conference on the industrial amd commercial use of energy, August 2011. Copenhagen, Denmark

Williams JA (2004) The impact of motivating factors on implementation of ISO 9001: 2000 registration process. Manage Res News 27(1/2):74–84

Wulandari M, Laskurain I, Casadesús M, Heras-Saizarbitoria I (2014) Impact of ISO 50001 standard: an empirical study. Documenta Universitaria, Girona

Sustainability in Iran's Road Transport Sector: Evaluating Strategies and Policies

Afsaneh Moradi, Seyed Reza Hejazi and Jahangir Yadollahi Farsi

Abstract Sustainable development is becoming increasingly important in the development of business strategies. Consequently, transport activities come under particular attention due to its significant role in economic development process and its impacts on the environment. Achieving sustainability in transport sector becomes more challenging given the growing demand for any kind of transport activities. The main aim of this paper is to explore the Iranian transport sector's sustainability barriers, and evaluate the efficiency of implemented policies and strategies in order to identify the potentials of improvement. For this purpose, an overview of the country and its transport sector in terms of energy consumption and pollutant emissions, which are two main recognized sustainability issues in Iran, will be discussed at first. Then, sustainability barriers are identified. To that end, the case study approach is used, and data are collected through documents analysis and interviewing transport planning and environment management experts. Then, the detected policies and strategies and their efficiency are discussed using the view point of transport and environment management experts by Delphi technique. The research revealed that the growing rate and the age of road vehicle fleet, outdated manufacturing standards, wasteful consumption patterns, traffic and inefficiency of public transport system are the main obstacles in the way of achieving a sustainable transport system in Iran. In addition, according to the viewpoint of experts, renovating deteriorated passenger cars, obligatory vehicle inspections, expanding subway system in metropolitans, renovating public transport vehicles and motivating private sector for investment in public transport were policies who considered having better performance. It is concluded that a sustainable transport system in Iran will be achieved only in case of changing the vehicle production and fuel consumption patterns. Also passenger travel and freight distribution should be shifted to other transport mode.

A. Moradi (✉) · S.R. Hejazi · J.Y. Farsi
Entrepreneurship Faculty, University of Tehran, Tehran, Iran
e-mail: a.moradi@ut.ac.ir

S.R. Hejazi
e-mail: rehejazi@ut.ac.ir

J.Y. Farsi
e-mail: jfarsi@ut.ac.ir

A. Chiarini (ed.), *Sustainable Operations Management*,
Measuring Operations Performance, DOI 10.1007/978-3-319-14002-5_10

Keywords Sustainability · Road transport · Strategy · Pollution · Emission · Evaluation

1 Introduction

Supply and proper use of energy is one of the most important issues that significantly influence sustainable development goals. For this goal, resource conservation and management, administrative use of new and renewable energies and reducing the environmental impacts of energy consumption is very important.

Transport sector has major share of energy consumption in any country and it has a key role in development process. Developing countries are facing unprecedented challenges towards sustainable societies in the sense that they have to balance economic growth and environmental consideration even though they are not major contributors to environmental loads (Fujiwara and Zhang 2005). These countries face a high rate of industrialization in a short period of time, as a consequence the demand for freight and passenger transport grows with an accelerated pace while the existing infrastructures are not able to meet the growing needs of development process. For this goal, policy makers may not pay enough attention to the context of sustainable development considering that achieving sustainability goals needs comprehensive planning in all aspects. In a developing country like Iran, current patterns of transport lead to unexpected environmental, social and economic costs. In this regard, some problems have been raised, including increased energy and fuel consumptions, road traffic, road safety, emissions and pollution and health impacts of road transport. Studies indicate that the environmental and social costs, in terms of local air pollutants, traffic accidents and congestion, can add up to nearly or over 10 % of a region or country's GDP (UNEP 2011; Creutzig and He 2009).

Transport sector's planning is one of the main one contributing in green growth agendas. According to Perkins transport sector's relevance for sustainable development has two main reasons: first, transport has major environmental impacts in terms of greenhouse gas emissions, local air emissions and noise. And managing congestion more effectively is part of the broader agenda for more sustainable development and better use of resources invested in infrastructure. Second, a large part of public expenditure to stimulate green growth is directed at transport sector industries. This concerns most notably alternative vehicles, and particularly electric cars, a key part of strategies to decarbonize transport.

In Iran a series of development plans has been defined for a period of 5 years each. Just after the end of the war, in 1989, government's strategies and policies in favor of a liberalized economy were formed in a series of new five-year development plans (FYDP). In recent years, policy makers try to face the challenge of economic growth and sustainability by modifying consumption patterns, and for transport sector some solutions are related to: improving infrastructure, developing appropriate and alternative modes of transport, upgrading and modernizing vehicle fleet, solving traffic problems, promoting the use of public transport, using cleaner fuels and avoiding unnecessary trips.

This paper aims at discussing sustainability barriers in Iran's road transport sector. Firstly, the current situation and emerging trends will be explained by the help of statistics and secondary sources analysis. The statistics and information are collected from national and international databases, such as World Bank, Iran's national accounts, annual transport books and reports, energy balance sheet etc. Secondly, semi structured interviews are conducted to collect information about the existing problems, plans and policies. The research revealed that there are two main critical issues in Iran's transport sector in the way of achieving sustainability goals:

1. The high share of energy and especially petroleum products consumptions in transport sector;
2. The high rate of air pollutant and CO_2 emissions from transport activities.

The paper is structured as follow: next section is related to a literature analysis about sustainability in transport sector and the country's profile monitored in two aspects; fuel consumption and pollution emission; after the methodology, the research observation and findings will be described based on transport and environment planning expert's opinions about the main challenges and problems which make difficulties for achieving sustainable development goals in transport sector. After that the effectiveness of policies and decisions will be evaluated with the application of Delphi method. At the end the policies are sorted based on the obtained scores and the reasons for the worst performance of policies which had the lowest scores will be explored.

2 Literature Review

2.1 Sustainability in Transport Sector

The term, sustainable development, gained the attention of international society in Our Common Future, which also known as the Brundtland report, published by the World Commission on Environment and Development in 1987, although sustainability issue had been noticed by some scholar before (Solow 1974; Hartwick 1977; Solow 1986). According to the World Commission on Environment and Development sustainable development is one that meets the needs of the present without compromising the ability of future generations to meet their own needs.

Sustainable development has proven an enduring and compelling concept because it points policy in a clear, intuitive direction, yet is flexible enough to adapt to emerging new issues, technological and economic conditions, and social aspirations. It is appealing to advocates and scholars alike because it implies a systemic view of economy and ecology, and requires comprehensive solutions that protect the interests of future generations. Some scholars who have written about sustainable development believe that in order to achieve more sustainable development patterns, the economic approach should be shifted to the environment and social problem solving approach (e.g. Jacobs 1991; Common 1995; Roodman 1998). Daly defines sustainable development as one that satisfies three basic conditions:

(a) Use of renewable resources does not exceed the rates of regeneration;
(b) Use of non-renewable resources do not exceed the rate at which sustainable renewable substitutes are developed; and
(c) Rates of pollution emission do not exceed the assimilative capacity of the environment.

Transport sector has proven to be particularly difficult territory for the advancement of sustainable development policy. Transportation is a complex and porous social, technical, and economic system, difficult to address comprehensively. To the extent that policy guidance has been developed to address sustainability issues in general, it usually has only touched on a fraction of the myriad ways that transportation is integrated into larger systems of human activity. Meanwhile, current trends are not encouraging.

Transport sector has a main part of economic activities. Transportation plays a significant role in all economic activities from the beginning of production process to delivering final products to the market. It serves increasingly as a prerequisite to maintaining and developing the productivity of other sectors of society. In developed countries, special attention is being given to the sustainability of current and emerging land use and transportation patterns. This focus reflects both the significant impacts that current patterns of transportation have on the environment and the complex interactions between transportation, land use, and activity systems. In this context, sustainable transportation is seen as transportation that meets mobility needs while also preserving and enhancing human and ecosystem health, economic progress, and social justice now and for the future. Planning for sustainable development aims to attain all three objectives simultaneously and in a just manner, considering access as well as mobility in the process (Deakin 2003).

A more sustainable approach to transportation will lead to less environmental damage before it occurs rather than as a reaction to the damage that results from vehicle emissions (Fergusson and Skinner 1991). There are two important aspects of environmental damage: the first is CO_2 emissions that cause climate change and the second is other emissions that contribute to air pollution and subsequent health effects. These two problems are exacerbated by the increasing trend for private vehicles. The trend calls for more vehicles on the road which means more vehicle emissions and thus more deleterious environmental effects. If public transportation is used more frequently, this can serve as a way to reduce the number of vehicles on the road and subsequently reduce emissions.

Considering co-benefits of sustainable transport and especially carbon mitigation policies, it can be argued that carbon mitigating policies and in general climate change policies have both side effects with sustainable transport planning. Many authors stimulate consideration of low-carbon, climate-resilient transport strategies and infrastructure planning from a "co-benefits" perspective. This enables the climate agenda to advance through synergies with more prominent policy goals since sustainable transport can generate social, economic and environmental co-benefits (Bollen et al. 2009a, b; Corfee-Morlot et al. 2009; Leather 2009; Viguié 2011; GIZ 2012; Zusman et al. 2012).

According to OECD, GHG emissions from transport sector will have to be significantly reduced (OECD 2009). Looking at the global trends, transport is the second largest contributor to global GHG emissions, causing 23 % of global carbon dioxide (CO_2) emissions from fossil-fuel combustion and about 15 % of global GHG emissions (Ang and Marchal 2013). In the absence of new policies, transport-related global CO_2 emissions are expected to double between 2010 and 2050, largely driven by emerging and developing countries (OECD 2012; IEA 2013).

Reducing emissions from transport sector is one of the main challenges that developing countries face in the way of achieving sustainable development goals. Generally transport demand has dramatically been increasing for many years throughout the world. Traffic levels are doubled in many car-depended countries in the period of 1975–1995 (Banister et al. 2000 and Banister 2000).

In Iran road transport has a considerable share of GHG and other pollutants in comparison with other subsectors as it's responsible for 92 % of NOx, 85 % SO_2, 99 % CO and 93 % CO_2 emissions, these statistics shows the need for urgent planning in order to limit the side effects of the growing rate of industrialization and urbanization in order to achieve sustainability in road transport sector. In addition, considering that the economic activities are highly depended to transport activities and transport sector is a main contributor of economic growth, it is forecasted that transport demand will be increased in future. Taking into account that Iran is an example of those countries that face several environmental and economic growth problems as the consequences of rapid urbanization the need for more studies in order to choose the most effective policies and the appropriate scales and instruments for measuring policy performances is emphasized.

2.2 Country Profile

Iran is a developing country located in a unique geopolitical region of south-west Asia (Middle East).Comprising a land area of 1,648,195 km^2 (636,372 sq mi), it is the second-largest nation in the Middle-Eastand the 18th-largest in the world; with 78.4 million inhabitants, Iran is the world's 17th most populous nation (Encyclopedia Britannica 2012).

According to the International Monetary Fund, Iran is a "transition economy", changing from a planned to a market economy (Jbili et al. 2007). Although the country is one of the world's major oil and gas producers, the oil sector accounts only for 25 % of GDP. The largest contributor of GDP is services sector (around 50 % of total output) (CBI 2012).

Transport in Iran is mainly road based. Road transport has the main share of passenger and freight displacement in Iran and is considered as the dominant mode in Iran's transportation sector with a share of 7 % in Iran's GDP and share of 90.5 % in passenger transport in 2011 (National Account of the Central Bank of Iran) Despite the major importance of this sector of transport in Iran, little researches have been done in this sector. The total road network length in Iran at 2011 was 229,057 km according to World Bank data that 74.3 % of them are paved.

National railway systems have often been used to reduce congestion and pollution, provide access to remote and small communities, or to support economic development and trade (OECD/ITF 2008) but in Iran the share of railway system in passenger and freight transport in negligible, government plans to transport 3.5 % of the passenger volume and 8.5 % of the freight volume by rail. The railway network expands by about 500 km per year according to the Ministry of roads and transportation (Abbaszadeh et al. 2010).

2.2.1 Energy Production and Consumption Trends

Iran has the fourth largest oil reserves and the world's second largest natural gas reserves (BP Statistical Review 2013; EIA 2013). The Iranian economy is highly depended on natural resources like oil and gas; the country was on a fast track before the global economic crisis erupted in 2008, averaging near 7 %. During this pre-crisis period, Iran's economic growth has been robust and its external position had strengthened on the back of rising international oil prices and pro-cyclical policies. The economy more than doubled from 2003–2004 till the end of 2008–2009 reaching USD 350 billion (eco trade and development bank 2014). After the economic crisis the global demand for oil and its by-products dropped, so the country lost an important part of its income, along this, the political situation intensified this reduction by reducing the country's supply share. Iran's oil production has substantially declined over the past few years, and natural gas production growth has slowed. International sanctions have profoundly affected Iran's energy sector (EIA 2014).

Natural gas and oil accounted for almost all (98 %) of Iran's total primary energy consumption in 2012, with marginal contributions from coal, hydropower, nuclear, and non-hydro renewable (Fig. 1). Iran's primary energy consumption has grown by more than 50 % over the past 10 years (EIA 2014). As it can be seen in this figure the natural resources share in energy consumption is about 98 % while the share of renewable energies is really negligible (less than 1 %), considering the world average of 67 % for natural resources and 3.4 % for renewable energies it can be argued that the first challenge in the way of achieving sustainable development goals for Iranian government it to limit the use of natural resources.

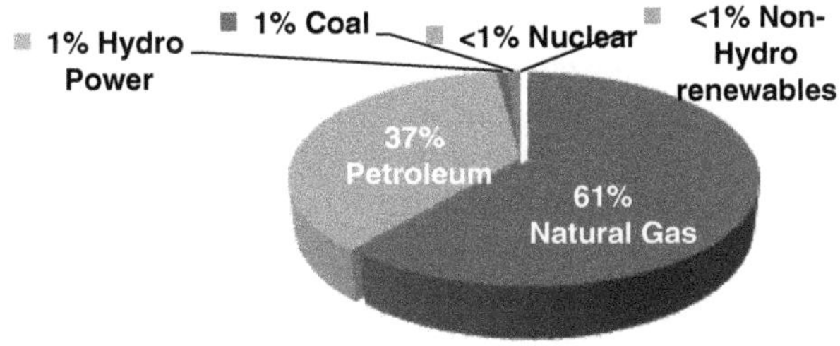

Fig. 1 Iran's energy consumption (EIA 2014)

Table 1 GHGs and air pollutant emissions caused by energy production and consumption (tons)

Year	NOx	SO_2	SO_3	CO	SPM	CO_2	CH_4	N_2O
2007	1,378,957	1,424,973	14,951	8,456,502	367,239	492,264,957	43,001	11,525
2008	1,808,553	1,598,617	15,853	8,973,628	386,757	523,293,617	46,214	12,101
2009	1,836,265	1,678,078	16,953	8,651,070	393,399	538,527,894	50,314	12,147
2010	1,805,823	1,355,656	13,896	8,201,223	655,600	532,324,843	52,236	11,967
2011	1,843,831	1,425,800	13,471	8,033,989	393,055	547,014,571	53,307	11,283
2012	1,861,374	1,540,500	13,636	8,523,752	400,983	556,866,442	57,049	11,609

Source Iran's energy balance sheet 2012

2.2.2 Pollution and Emission Trends

Energy production and consumption have negative impacts on environment. Air pollution is the most critical energy-related problem in Iran's big cities. According to world health organization, four cities in Iran make the top 10 list of most polluted cities in the world. Table 1 shows the trends of GHGs and air pollutant emissions caused by energy production and consumption between the years 2007 and 2012. It's clear that the share of CO_2 is more than any other pollutant.

The rapid urbanization, population increase, growth of the industry and transportation sectors, and the inefficient consumption patterns have increased concerns with regard to the emission of pollution. The urban environment of Iran is becoming increasingly polluted, with adverse impacts on the health, welfare and productivity of the population. Pollution in Tehran, where 20 % of Iran's population lives, has well exceeded safe levels (Masjedi et al 1998; Asgariet et al 1998 and EIA 2000). Rural areas are threatened by over-exploitation, soil erosion, pollution from agrochemicals and inappropriate land use and management practices (Rostamihozori 2002).

3 Research Methodology

This research is an empirical exploratory case study which is used to explore those situations in which the intervention being evaluated has no clear, single set of outcomes (Yin 2003).; the main aim of research is to look for sustainability problems in Iran's road transport sector. After explaining the country's situation and emerging trends with statistics and secondary resources, 21 semi structured interviews conducted to collect necessary information about the existing problems, resolving plans and policies.

The statistics and information are collected from national and international databases, such as World Bank, Iran's national accounts, annual transport books and reports; Energy balance sheet, etc. The interviewees have been chosen by the selective sampling method from transport planning experts and managers. The interviewees asked to give their idea about transport sector's barriers for achieving sustainability goals, current and future policies or plans for resolving those problems. The answers have been analyzed and the problems that have been mentioned by at least 12 interviewees (more than 50 % of interviewees) considered as the country's main challengers for achieving sustainable transport. The identified problems have been examined by supporting statistical data. Then the effectiveness of policies and plans has been analyzed based on Delphi method by a panel of experts who asked to answer the questionnaires in two rounds. They have been asked to weight the list identified policies (in order to find out which policy is preferred) and give their opinion about the effectiveness of the selected policies and plans by giving a number between one and ten according to the efficiency of each policy. At the end the average score is calculated for each policy and the policies and been sorted according to the obtained score.

4 Observation and Discussions

4.1 Sustainability Barriers in Road Transport Sector

According to transport planning experts in Iran, the rapid rate of urbanization, growing rate of petroleum products consumption as a consequence of growing number of vehicles in the roads, the age of vehicle fleet, manufacturing vehicles with abolished fuel consumption standards, the inefficiency of public transportation system, the low price of petroleum products are among the obstacles of achieving sustainable transport system in Iran. Here the problems will be explained with the support of statistical data and other researches achievements:

1. **The growing rate of road vehicle fleet**
   In Iran, about 90 % of passenger and freight are moved by road transport. General road transport fleet including truck, bus, minibus and passenger cars is the key element of the transporting passengers and cargo on the roads. Between the years 2001 and 2011 the number of vehicle freight shows an annual growth rate of about 12.39 %whilst the average growth rate for passenger cars in the world is 2.9 % and for heavy duty vehicles is 4.52 %.
   Beside vehicle fleet, another problem is the number of motorcycles which came into the market without considering any environmental standard. The motorcycles fleet also rose with an accelerated pace of annual growth rate about 21.96 % between 2001 and 2011.

2. **Vehicle fleet age**
   In Iran vehicles are used more than their optimal life due to the high ratio of vehicle to fuel prices and this cause the vehicle fleet age raise (Salahi and Kashani 2008). From 1990s many countries around the world took incentive or mandatory policies for phasing out old vehicles. In Iran, there were incentive strategies for replacing old passenger cars by the new ones after the year 2000. The average age of buses, trucks, taxi and personal vehicles was 22,19 and 15 years in 2005, and by removal of 758 trucks, 5,502 buses and minibuses and 147,475 taxi and the obligation from government along with loan and financial aid, the average age of these vehicles have reached 19, 13 and 15 years. But there is a need for more investments for replacing medium and heavy duty vehicles, as the majority of those vehicles are still above their optimum lifetime. The lack of investment for medium and heavy duty vehicles resulted in reducing the effectiveness of renovating light duty vehicle fleet by increasing the share of old heavy duty vehicles in fuel consumption and pollution emissions. The statistics confirms this statement by representing that 80 % of light duty vehicles has the ages less than 10 years, while 60 % of medium duty passenger vehicles and heavy duty trucks still working with more than 20 years old. Considering that the amount of pollutants emitted by old vehicles is many times more than a normal one, the necessity of renewing medium and heavy duty vehicle fleet could be emphasized.
3. **Manufacturing vehicles with abandoned or outdated fuel consumption standards**
   Several EU transport emission-related policy measures, such as fuel quality and ever tighter vehicle emission 'Euro' standards for certain new vehicles, have significantly reduced emissions from the transport sector (EEA 2010). In Iran most of the manufactured vehicles have been out of Euro emission standard scopes and those which were in accordance with Euro emission standards, couldn't obtain more than Euro 4 standard. Considering the combined fuel consumption standards, vehicles in Iran are generally consume more than 6, 5 l, and even sometimes about 14 l per 100 km in full circulation. Although there are obligatory regulations for technical inspection of vehicles and heavy fines for those who do not obey, but the vehicle manufacturing companies in Iran couldn't or didn't want to adapt themselves with global trends.
4. **Wasteful consumption patterns due to low fuel prices**
   The energy market in Iran is a monopoly. The government, as the only supplier, sets the prices of energy carriers. The government has historically set prices low. Taking into account its average price in Persian Gulf to be almost 4 US$, it can be concluded that it was being supported by government subsidiary (Rasouli Rasouli 2009).Generally speaking; the government froze energy carriers' prices from 1978 to 1996. By 1996, given the inflationary state, the price freeze depreciated energy prices. The resulting distortion in energy prices in Iran has had a two-fold effect:

(a) Energy prices are generally far below the competitive global market price; and
(b) The relative prices of energy carriers are distorted with respect to each other.

In 1996 government started a gradual price reform on energy carriers (Sabetghadam 2006). The first phase of the reform was enacted in late 2010, and phase two was initiated in early 2014 (EIA 2014).This reform aimed at correcting the relative prices of oil products and their relationship to the general consumer price index. However because of the gradual and insufficient increase in the price of energy, the policy has not been effective (Sabetghadam 2006).

5. **Traffic**

   In many countries, the exclusive concentration on a simple extension of the road based transportation network led to growing individual car traffic hence to additional demand for roads and striking external effects of transport (Arndt and Doge 2013).

   Traffic is the main source of fuel consumption and pollution emissions especially in urban areas. Most of the Iranians prefer to use private cars due to inefficiency of public transport system. Private cars number shows the annual growth rate of 12.06 between the years 2001 and 2011, while the annual growth rate for taxies, buses and mini buses was 10.32, 5.72 and 1.30 respectively. In addition stable traffic flows in big cities increases the fuel consumption and pollution emission in certain hours of the day.

6. **Inefficiency of public transport system**

   The annual growth rate for the use of public bus and mini buses in Iran has a minus trend. According to Transport and Energy Data Book, the annual growth rate for the use of private cars, taxies, buses and mini buses in Iran between the years 2001 and 2011 was respectively: 12.06, 11.38, −2.43 and −5.22 %. This statistics shows that people do not have tendency to use public transport system. This may be because of inadequate public transport vehicles and infrastructures, time consuming journeys with public transport and traffic conditions in big cities. Figure 2 shows the trends of private car use in comparison with public transport vehicles. As it is clearly shown in Fig. 2, people in Iran prefer to use private cars instead of public transport, although public transport is much cheaper than other countries in the world.

4.2 The Main Challenges for Achieving Sustainable Development in Transport Sector

The problems described above have some macro-level consequences which cause two main challenges for sustainable transport system in the country. These challenges are:

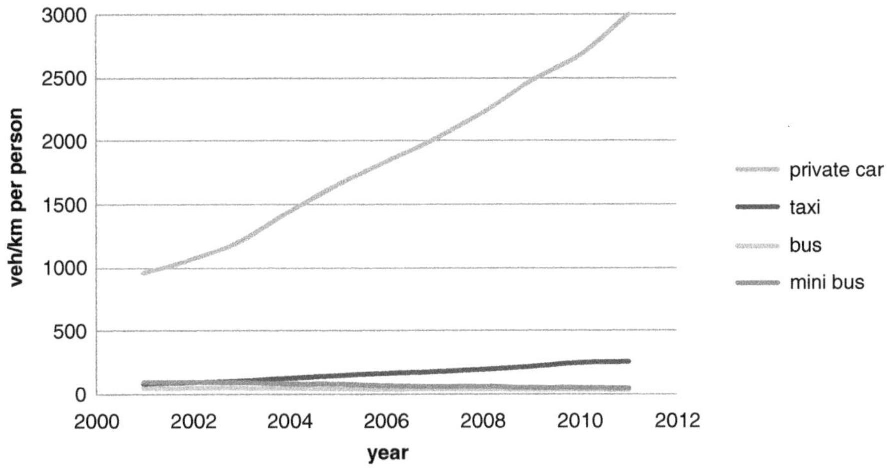

Fig. 2 Trends of private car use in comparison with general public transport vehicles. *Source* transport and energy data book

1. The high share of energy and especially petroleum products consumptions in transport sector;
2. The high rate of air pollutant and CO_2 emissions from transport activities.

The challenges trends can be monitored by the statistical data obtained from national resources. The government took some policies and strategies for resolving these problems. In the next part, the policies and their efficiency evaluated by the transport planning experts will be discussed.

4.2.1 Transport Sector's Higher Share of Energy Consumption

Transport sector is one of the foremost primary energy consumers of fossil fuels, especially petroleum products, and is a major contributor to environmental degradation (Wen and Yan 2010; Serrano et al. 2012). Transport sector emissions are a main component of sustainability problems like air pollution and climate change.

As previously mentioned, transport sector is one of the main objective sectors in planning for sustainable development. In transport sector the main energy consuming subsectors are road, air and marine transport (Dincer et al. 2004). The growing number of private cars, buses and motorcycles on the road subsector on one side, and improving the airplane, aircraft and cargo ship's quality in air and marine subsectors on the other side, increases the energy consumption of this sector (Wen and Yan 2010). The consequence of transport sector development is the depletion of natural resources, and the increased greenhouse gases emissions, which attracts the attention of both the scientific community and policy makers (Ozan et al. 2011).

In Iran road transport accounts for about at least 25 % of total energy consumption and has the highest share of consuming petroleum products. Table 2

Table 2 Energy consumption in transport and other sectors, 2001–2010

Year	Transport share of energy consumption (%)	Petroleum products consumption in transport sector (%)
2001	28.5	99.98
2002	28.6	99.98
2003	28.7	99.98
2004	28.2	99.75
2005	28.2	99.22
2006	27.1	98.75
2007	24.5	97.49
2008	25.3	95.83
2009	26.5	92.93
2010	26.1	88.29

Source Electricity and energy general planning office, ministry of energy, Iran

shows the energy and petroleum products consumption in transport sector compared with all other sectors between the years 2001 and 2010.

4.2.2 Pollution and Emissions in Iran's Transport Sector

In Iran transport sector has the main share of any gas emission between all energy consumers. It should be emphasized that transport sector is the second largest contributor to CO_2 emissions after power plant sector. As it's shown in Table 4 again CO_2 emissions in road transport sector is more than any other pollutant (Table 3).

Iran is one of the highest carbon emission-intensive countries in the world. According to World Bank data, Iran ranked 10 for CO_2 emissions in the world and in the most recent data published by IEA at 2013, the country become 8th CO_2 emitting country in the world. Beside this, at 2011 the CO_2 emission per capita in

Table 3 The percentage of GHG and other pollutant emissions in energy producing and consuming sectors (%), •: data are not available

	NOx	SO_2	SO_3	CO	SPM	CO_2	CH_4	N_2O
Total energy consuming sectors								
Residential, commercial and public sector	5.76	3.49	4.50	0.53	2.62	22.92	7.05	4.46
Industry	9.03	13.15	21.97	0.17	4.30	16.99	3.64	2.40
Transport	48.12	25.83	31.73	97.22	78.48	23.49	80.07	51.08
Agriculture	3.27	4.06	2.79	0.17	6.64	2.26	1.14	35.65
Energy producing sectors								
Refineries	•	•	•	•	•	2.98	0.62	0.39
Power plants	33.81	53.46	39.01	1.90	7.97	31.37	7.49	6.01

Source Iran's energy balance sheet 2012

Table 4 GHGs and air pollutant emissions in Iran's road transport sector (tons)

Year	NOx	SO_2	SO_3	CO	SPM	CO_2	CH_4	N_2O
2007	780,647	325,172	3,456	8,268,940	258,318	106,983,905	32,416	5,159
2008	816,610	340,749	3,623	8,619,911	270,677	113,235,805	35,293	5,389
2009	807,245	340,992	3,642	8,312,710	270,761	115,820,357	39,526	5,364
2010	801,994	346,931	3,737	7,849,053	275,266	117,470,093	41,575	5,254
2011	798,551	347,584	3,752	7,707,863	275,729	117,951,801	42,390	5,212
2012	828,475	356,394	3,831	8,208,263	282,824	123,858,601	45,595	5,430

Source Iran's energy balance sheet 2012

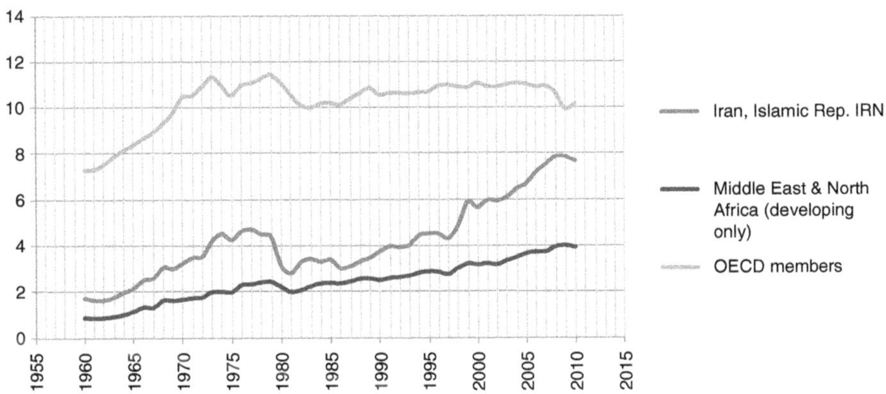

Fig. 3 The rate of CO_2 emission trends in Iran, with Middle East developing and OECD countries. *Source* World Bank development indicators

Iran was 7.32 tons per person which is a considerable amount. The CO_2 emissions in Iran dramatically rose in 10 years; between 1998 and 2008 the country's CO_2 emission amount get almost doubled (Fig. 3). Iran has the highest rate of CO_2 emissions in the region and the country's share is more than all developing countries in Middle East and North Africa.

4.3 The Existing Policies

Iranian government has special policies for improving transport sector efficiency, especially in line with the main challenges mentioned above; energy consumption and pollutant emissions. Different organizations and ministries asked to cooperate for solving the problem of air pollution which is now a critical problem in the country. Also some organizations established especially for these goals like the Fuel conservation organization (IFCO) that was established at 2000 to help promote energy efficiency and conservation. IFCO studies energy consumption standards and conducts research on ways to apply such standards.

Table 5 The main policies for facing the challenge of high air pollution and fuel consumption in Iran

Main goal	Existing policies
Transport supply management	Transport integration
	Shifting freight and passenger transport demand to other transport modes
	E-governance
	Improving roads, highways and other transport infrastructures
Fuel consumption management	Vehicle technology improvement
	Gradual elimination of fuel subsidiaries
	Using alternative fuels like CNG
	Renovating deteriorated cars
	Duty-free fuel efficient and electric and hybrid vehicles
Air pollution reduction	Monitoring air pollution in cities
	Vehicle technical inspection
	Electric motorcycles
Traffic management	Traffic restricted areas in metropolitans
	Expanding subway system in big cities
	renovating public transport vehicle fleet
	motivating private sector to invest in public transport
	Bus rapid transit lines
	Parking supply management

The existing policies include different plans and strategies from fuel consumption management to transport modal shift strategies and vehicle maintenance and mandatory inspection regulations. The main strategies mentioned by the interviewees are classified and summarized in the Table 5 which was given to Delphi panel experts to be weighted.

They have been asked to weight the list identified policies (in order to find out which policy is preferred) and give their opinion about the effectiveness of the selected policies and plans by giving a number between one and ten according to the efficiency of each policy. At the end the average score is calculated for each policy and the policy according to weighted average equation:

$$\mathrm{Ps} = \frac{\sum_{i=1}^{n} W \times E}{\sum W}$$

where in it:

Ps Policy Score

n the number of Delphi panel participants

E Policy effectiveness

W Policy weight

Table 6 The performance of existing policies adopted for reduction of air pollution and fuel consumption in Iran's transport sector

Policy	Policy scores
Renovating deteriorated cars	7.09
Vehicle technical inspection	7.05
Expanding subway system in big cities	6.64
renovating public transport vehicle fleet	6.54
motivating private sector to invest in public transport	6.52
Using alternative fuels like CNG	6.35
Bus rapid transit lines	5.88
Duty-free fuel efficient and electric and hybrid vehicles	5.30
Parking supply management	4.94
Electric motorcycles	4.91
Monitoring air pollution in cities	4.85
Improving roads, highways and other transport infrastructures	4.77
E-governance	4.50
Transport integration	4.20
Shifting freight and passenger transport demand to other transport modes	3.85
Traffic restricted areas in metropolitans	3.55
Gradual elimination of fuel subsidiaries	3.24
Vehicle technology improvement	3.05

The policy score obtained in this formula shows that which policies have been more effective in Iran. It is not necessarily means that the policies with higher score are better, but those policies had better performance in achieving sustainable transport system goals in Iran. Table 6 shows the policies sorted based on the score obtained in weighted average equation.

As it as shown in Table 6, according to Delphi panel experts; renovating passenger cars, vehicle technical inspection and expanding subway system in metropolitans were the most effective policies in reducing air pollution and fuel consumption in Iran. The experts believed that the vehicle technology improvements, elimination of fuel subsidiaries and traffic restricted areas were not successful policies. This doesn't mean that these policies shouldn't be applied any more, but what is evaluated in this paper is the policy performance in Iran during the 10 years between 2001 and 2011 and indeed there were some reasons for limiting the performance of the strategies and policies.

So at the second round of Delphi method the results of the analysis has been given to the panel members and they asked to give their opinion about the reasons of ineffectiveness of the last five policies listed in Table 6.

The main reasons that prevent the policies to be effective are listed in Table 7.

Table 7 The main reasons of policy ineffectiveness

Policy	The main reasons for low performance
Transport integration	Last efforts for modifying land use patterns in Tehran and some other big cities was not successful
	Residential areas around cities do not have suitable access to the city centers
	Low taxi fares caused people do not have the tendency to use other means of public transport like bus and subway system
Shifting freight and passenger transport demand to other transport modes	The rail network coverage in the country is not suitable
	Air transport does not have enough capacities to handle the burden of more passenger or freight transport
Traffic restricted areas in metropolitans	there is a possibility to buy the traffic authorization to enter the restricted areas, the process is not preventive
Gradual elimination of fuel subsidiaries	Since the consumption behavior and the vehicle consumption amounts are not modified this policy wouldn't be effective
Vehicle technology improvement	The improvement in vehicle technology is not according to the updated standards in the world
	Due to the high prices of passenger cars, customer are reluctant to pay more for advanced technologies

5 Conclusion

Achieving sustainable transport system in developing countries is a challenging process. As these countries experience high urbanization rates in a short time, the transport infrastructures are not adapted to the pace of urbanization, on the other side, the growing number of economic and business activities are highly depended to the transport activities. As a result, transport system grows without considering the structural issues. Iran is an example of those countries that face several environmental and economic growth problems as the consequences of rapid urbanization.

This research aimed at identifying sustainability barriers in Iran and evaluating the implemented policies for resolving those problems, the research revealed that the high growth rate of vehicle fleet and the fleet age, outdated production standards, wasteful consumption patterns, traffic and inefficiency of public transport system are the main problems in the way of achieving sustainability goals in transport sector. The main macro level consequences of those problems are:

1. The high share of energy and especially petroleum products consumptions in transport sector;
2. The high rate of air pollutant and CO_2 emissions from transport activities.

Monitoring the trends of those macro-level challenges showed that the country is in a critical situation now and there is an urgent need for a comprehensive planning in order to reduce the consumption of natural resources and air pollution emissions.

The main policies implemented by the government in the studied period (2001–2011) identified by interviewing transport and environment planning experts and managers. Then the policies were given to the panel of experts they asked to weight the policies according to their importance and score the performance of those policies. The weighted average for each policy was calculated based on the scores was given by the experts, and then the policies were sorted according to the obtained score. The list of policies and the main aim of applying those policies are presented in Table 5. The sorted list of policies based on policy scores can be seen in Table 6.

Although all policies listed in Tables 5 could contribute in making transport more sustainable, at least they can contribute to solve one of sustainability obstacles that were mentioned in Sect. 4. However, if the policies or strategies could not be managed properly like the elimination of fuel subsidiaries, their effect will be limited as the transport and environment sectors experts claimed at the second round of Delphi questionnaire. Consequently, there should be predefined metrics and objective goals to measure the policy effectiveness while in Iran there is no clear objective for evaluating the effectiveness of implemented policies. Another problem is that some of those policies implemented in a certain period of time and after that they have been abandoned or excluded due to the changes in the top management level as a consequence of government changes.

Ultimately a sustainable transport system in Iran will be achieved only in case of changing the vehicle production and fuel consumption patterns. Also passenger travel and freight distribution should be shifted to other transport mode. The research showed that if the policy chosen based on the comprehensive study of future effects and well managed during the time of the application, they could be introduced as part of broader strategies. Ultimately, it is likely that only through such a broad approach, alongside the policy instruments above, the transport sector will be able to achieve sustainable development objectives.

The practical conclusion of this research consists in the necessity to optimize public transport fleet and infrastructure, choosing the best policies for limiting the use of private cars, reforming vehicle manufacturing industry and Selecting policies and strategies based on the scientific decision making process instead of choosing available possibilities. Government and public administration has to play a key role in this respect, thus the paper tries to underline the importance of choosing the effective policies and evaluating the effectiveness of applied policies and strategies. Another important issue is that people should be informed about the consequences of pollution and Indiscriminate consumption of fuels. They should be motivated to voluntarily cooperate in implementing the strategies.

In Iran many researches had been done for studying the air pollution in Tehran metropolitan area (Asadollah-Fardi 2001; Abbaspour and Soltaninejad 2004; Moradi et al 2010; Torkian et al 2012), however more research is needed with the whole country vision considering that at least two or three other cities (like Ahwaz and Kermanshah) ranked among the most polluted cities in the world.

References

Abbaspour M, Soltaninejad A (2004) Design of an environmental assessment model on the effect of vehicle emission in greater Tehran on air pollution with economic sensitivity. Int J Environ Sci Technol 1(1): 27–38

Abbaszadeh SA et al (2010) Nonlinear site response evaluation procedure under the strong motion: a case study of Miyaneh Azarbayjan Sharghi Province Iran. Sci Res Essays 5(16):2257–2274

Ang G, Marchal V (2013) Mobilizing private investment in sustainable transport: the case of land-based passenger transport infrastructure. OECD environment working papers 56, OECD Publishing. doi:10.1787/5k46hjm8jpmv-en

Arndt WH, Döge N (2013) The local (public) transport plan as an approach to optimize urban public transport planning in Iran. Technische Universität Berlin Research report. ISBN 978-3-7983-2586-9

Asadollah-Fardi G (2001) Current situation of air pollution in Tehran with emphasis on District 12. First Kitakyushu initiative meeting, Kitakyushu

Asgari M, DuBois A et al (1998) Association of ambient air quality with children's lung function in urban and rural Iran. Arch Environ Health 53:222–230

Banister D (2000) Sustainable urban development and transport a Eurovision for 2020. Transp Rev 20(1):113–130

Banister D, Stead D, Steen P (2000) European transport policy and sustainable mobility. Spon Press, London

Bollen J et al (2009a) Co-benefits of climate change mitigation policies: literature review and new results. OECD Economics Department Working Papers 693. OECD Publishing doi:10.1787/224388684356

Bollen JC et al (2009b) Co-benefits of climate policy. PBL Report 500116005, Netherlands Environmental Assessment Agency (PBL)

Central bank of Islamic republic of Iran (2012) Annual national accounts and statistics data. Latest edition

Common M (1995) Sustainability and policy: limits to economics. Cambridge University Press, Cambridge

Corfee-Morlot J et al (2009) Cities climate change and multilevel governance. OECD Environment Working Papers 14. OECD Publishing doi:10.1787/220062444715

Creutzig F, He D (2009) Climate change mitigation and co-benefits of feasible transport demand policies in Beijing. Transp Res Part D Transp Environ 14(2):120–131

Deakin E (2003) Sustainable development and sustainable transportation: strategies for economic prosperity environmental quality and equity. Institute of Urban and Regional Development, University of California at Berkeley, Berkeley

Dincer I, Hussain MM, Al-Zaharnah I (2004) Energy and exergy utilization in transportation sector of Saudi Arabia. Appl Therm Eng 24:525–538

Energy Information Administration (2000) Iran; environmental issues. http://www.eia.doe.gov

Energy Information Administration (2014) Country analysis brief: Iran independent statistics and analysis

European Environment Agency (2010) Towards a resource-efficient transport system: (2009) indicators tracking transport and environment in the European Union. European Environment Agency Technical reports 2. Available at: http://www.eea.europa.eu/publications/towards-a-resource-efficient-transport-systems

Fergusson M, Skinner I (1991) Greening Transportation Environment. Res Libr Core 41(1):24

Fujiwara A, Zhang J (2005) Evaluating sustainability of urban development in developing countries incorporating dynamic cause-effect relationships over time. J Eastern Asia Soc Transp Stud 6:4349–4364

Gesellschaft für Internationale Zusammenarbeit (2012) Navigating transport NAMAs: practical handbook for the design and implementation of nationally appropriate mitigation actions (NAMAs) in the transport sector. Draft v0.2 transfer project towards climate-friendly transport technologies and measures deutsche (GIZ) GmbH http://www.transferproject.org/images/Handbook%20v0.2/ManualNTN_V2_final_master2.pdf

Hartwick JM (1977) Intergenerational equity and the investing of rents from exhaustible resources. Am Econ Rev 67(5):972–974

International Energy Agency (2013) Global land transport infrastructure requirements estimating road and railway infrastructure capacity and costs to 2050. Insights paper IEA/OECD Paris http://www.iea.org/publications/freepublications/publication/TransportInfrastructureInsights_FINAL_WEB.pdf

Iran. Encyclopedia Britannica (2012) Encyclopedia Britannica

Jacobs M (1991) The green economy: environment sustainable development and the politics of the future. Pluto Press, London

Jbili A, Kramarenko V, Bailén JM (2007) Islamic republic of Iran: managing the transition to a market economy. The International Monetary Fund. ISBN 978-1-58906-441-6. Retrieved 5 Feb 2011

Leather J (2009) Rethinking transport and climate change. ADB sustainable development. Working paper series No10. Asian development bank (ADB). Available at: http://www.adb.org/sites/default/files/pub/2009/ADB-WP10-Rethinking-Transport-Climate-Change.pdf

Masjedi M et al (1998) Survey of air pollutants in Tehran. J Med Coun I.R.I 4:155–62

Moradi G et al (2010) Investigation of photochemical smog episode in Tehran city. In: International conference on environment 2010 (ICENV 2010)

OECD (2009) Reducing transport GHG emissions: opportunities and costs. OECD Publishing/International Transport Forum Paris

OECD (2012) OECD environmental outlook to 2050: the consequences of inaction. OECD Publishing doi:10.1787/9789264122246-en

OECD/ITF (2008) Transport infrastructure investment: options for efficiency. OECD Publishing doi:10.1787/9789282101568-en

Ozan C, Haldenbilen S, Ceylan H (2011) Estimating emissions on vehicular traffic based on projected energy and transport demand on rural roads: policies for reducing air pollutant emissions and energy consumption. Energy Policy 39:2542–2549

Rasouli E (2009) A review of energy consumption in the transportation sector in Iran and policies for reduction in the energy consumption in this sector. In: 8th International conference on sustainable energy technologies. Aachen

Roodman DM (1998) The natural wealth of nations. WW Norton London, London

Rostamihozori N (2002) Development of energy and emission control strategies for Iran. Ph.D. dissertation, University of Karlsruhe, Karlsruhe

Sabetghadam M (2006) Energy and Sustainable Development in Iran. Report made by Institute for International Energy Studies, Helio International Sustainable Energy Watch

Salahi J, Kashani B (2008) Economic analysis for the replacement of old heavy-duty vehicles in Iran. J Econ Bus Res 1(2):35–47

Serrano LMV et al (2012) Performance study about biodiesel impact on buses engines using dynamometer tests and fleet consumption data. Energy Convers Manage 60:2–9

Solow RM (1974) Intergenerational equity and exhaustible resources. In: The review of economic studies. Symposium on the economics of exhaustible resources vol. 4, pp.1 29–45

Solow RM (1986) On the intergenerational allocation of natural resources. Scandinavian J Econ 88 (1):141–149

Torkian A et al (2012) Source apportionment of Tehran's air pollution by emissions inventory. In: Emission inventories—meeting the challenges posed by emerging global, national, regional and local air quality issue. USEPA, Tampa, pp 41–41, 13–16 Aug 2012

UNEP (2011) Towards a green economy: pathways to sustainable development and poverty eradication—a synthesis for policy makers. Available at: www.unep.org/greeneconomy France

Viguié V (2011) Modélisation des dynamiquesurbaines application à l'analyseéconomique du changementclimatique. Available at: http://www.centrecired.fr/IMG/pdf/resume_pour_site_web_francais-2.pdf

Wen XJ, Yan L (2010) Analysis of corporate behaviors on different energy saving emission reduction policies. Ebm Int Conf Eng Bus Manag 1–8:3338–3341

Yin RK (2003) Case study research: design and methods, 3rd edn. Sage, Thousand Oaks

Zusman E, Srinivasan A, Dhakal S (2012) Low carbon transport and co-benefits in Asia. In: Zusman E, Srinivasan A, Dhakal S (eds) Low carbon transport in Asia: strategies for optimizing co-benefits. Earth scans, Taylor & Francis, New York

Milton Keynes UK
Ingram Content Group UK Ltd.
UKHW022204241024
449909UK00003B/55

9 783319 140018